D0867085

FUTURE RICH

FUTURE RICH

The People, Companies, and Industries Creating America's Next Fortunes

by JACQUELINE THOMPSON

William Morrow and Company, Inc. New York

ALBRIGHT COLLEGE LIBRARY

Copyright © 1985 by Jacqueline Thompson

All rights reserved. No part of this book may be reproduced or utilized in any form or by any means, electronic or mechanical, including photocopying, recording or by any information storage and retrieval system, without permission in writing from the Publisher. Inquiries should be addressed to Permissions Department, William Morrow and Company, Inc., 105 Madison Ave., New York, N.Y. 10016.

Library of Congress Cataloging in Publication Data

Thompson, Jacqueline.
Future rich.

Bibliography: p.
Includes index.
1. Technological innovations—United States.
2. Wealth—United States. I. Title.
HC110.T4T47 1985 338.4′7 84-20590
ISBN 0-688-04039-X

Printed in the United States of America

First Edition

1 2 3 4 5 6 7 8 9 10

BOOK DESIGN BY PATTY LOWY

338.47
T473f

196791

For Dorothy and Philip
whose contribution
to the future
is embodied
in their children

Preface

Mr. and Mrs. Jones are awakened by a soft, persistent beep, followed by a monotone voice synthesizer that says, "Good morning, Frank and Anne. It's six o'clock. Time to wake up." The button on the electronic bedside console that Anne Jones pushes to silence the beeper also activates their venetian blinds, which will rise and fall for the rest of the day at the whim of a tiny, energy-conscious microprocessor.

As the sun streams into the room, Anne Jones jostles her husband to get up. In two hours he's due at the Zorac Auto Company's assembly plant, where he's responsible for the programming, maintenance, and repair of five hardworking robots. His robots, in tandem with other robot crews, perform 97 percent of the manual labor in the plant, assembling thirteen hundred cars each twenty-four-hour day without any coffee breaks. Anne Jones, on the other hand, works at home. She's a text editor for the Editorial Services Publishing Company forty miles away. At a specified hour each day, she receives raw manuscript copy from her employer via telephone. It is stored in a word processor until she is ready to sit down at the video display screen and begin work.

For the past few mornings, Anne Jones has found her two children huddled in front of the family television set in their pajamas playing Monopoly, an interactive videodisk game to which they've developed a virtual addiction. She isn't surprised. After all, the game is fun. When it's your turn, you get to drive your car through a cartoon town featuring the familiar streets and utilities of an antique Monopoly board.

But Anne Jones knows she won't find her children in the living room this morning because she locked the doors of the television cabinet with her signet ring before going to bed last night. The same electronic chip that is embedded in her ring activates all the locks in the house. Each lock contains an electronic scanning device that reads the code on the ring. Frank Jones has his chip on his watch-band.

That's not the only wonder of modern technology that Frank Jones wears on his wrist. Frank Jones is a diabetic and has to receive periodic insulin injections. He never liked giving himself the shots, but now that onerous task has been eliminated—by his watch! His digital wristwatch actually does quadruple duty: It tells the time . . . it is a miniature calculator . . . it acts as a diagnostic machine . . . and it performs the job of a registered nurse, giving him the vital injections.

A physician monitoring Frank Jones's body around the clock couldn't be more vigilant than his wristwatch. The watch comes equipped with a microfine probe that extracts blood for composition analysis. A tiny computer evaluates the blood sample, his sweat, and other body secretions and takes regular readings of his pulse rate and blood pressure.

Equally miraculous is the minuscule hypodermic syringe that automatically gives Frank Jones insulin injections whenever he needs them. The insulin is as pure as nature could make it because it was manufactured by a drug company using the latest bioengineering techniques. The gene machines in the company's factory may be expensive, but the raw materials used to make the insulin are not. They are simply sugar, salt, and the isolated DNA sequence that codes for insulin production.

Anne Jones puts her *Good Housekeeping* menu-planner disk into one of the family's three home computers to find out what she's supposed to serve for breakfast this morning. The video display screen informs her that blueberry pancakes are the order of the day. She instructs the computer to call up the recipe, then has the attached dot-matrix printer type out the recipe on a sheet of paper. She finds it easier to work from an old-fashioned piece of paper, although many of her friends prefer to read the instructions directly off the screen.

Anne Jones knows she has the ingredients on hand because the

Good Housekeeping menu planner also formulates a shopping list to go with each week's complement of recipes. Anne Jones does her grocery shopping at home via a home computer/cable TV hookup. The brand-name items in each product category available at her local supermarket are listed on the videoscreen and she merely types the item codes into the computer. Then she types in her bank checking account number to pay for her order. Within two hours, bags of groceries are delivered to her doorstep.

By seven-fifteen, Frank Jones is settling behind the wheel of his car —fueled by electricity—for the drive to the factory. The thirty-minute commute is one of the pleasantest times of the day for him, since several tiny microprocessors make most of the driving decisions. The speedometer and capacitor are just a few of the parts that are electronically controlled. But the best feature of all, in Frank's opinion, is the microprocessor that computes the speed of the car traveling in front of his and assesses whether the two vehicles are being driven at a safe distance from each other. On foggy or overcast days, the headlights even come on automatically when it gets too dark.

Anne and Frank Jones are not characters in a science-fiction fantasy. They're fictitious people, all right, but their environment is very real —or will be reality a little more than a decade from now. It's the same environment we will all be experiencing by the year 2000. For by that time, the United States—along with the other advanced nations of the world—will be ending a three-decade-long, wrenching transition from a fossil-fuel-based, mechanical society to an electronics-based, fully computerized society.

It's been called the "postindustrial state," the "electronics age," the "third wave," the "information society," and the "second industrial revolution." Whereas the old industrial system raped and pillaged and destroyed the landscape, the new order will show more respect for nature's bounty. Its energy sources will be largely renewable, its primary industries will be relatively clean and energy misers, and its ethic will call for recycling raw materials. Today, as I sit and write, Japan is already halfway there.

Daniel Bell, the Harvard University sociologist who popularized the term "postindustrial society," describes turn-of-the-twenty-first-century America as a country that will emphasize human (that is, education, health, and social services) and professional (computing, systems analysis, and scientific research and development) occupations and services.

In the new age, brains will replace brawn as the prime force in society. People will be making more decisions—and more informed decisions—while electronic-powered machines carry out their commands.

The blue-collar worker once symbolized the predominant form of labor during the industrial smokestack phase of American history. But the white-collar worker will symbolize the types of greaseless employment common in postindustrial society. Jobs that require musclepower will be performed not by sweating men nor straining animals but by machine replicas of them. These robots will have many of the same sensory capabilities as people.

Boring, repetitive office jobs, whether they involve numbers or words, will also be relegated to sophisticated machines—mainframe, mini-, and microcomputers as well as word processors. Less paper will be shuffled on desks, among desks, and among companies because computers will serve as in-boxes, file cabinets, dictation and composing machines, and electronic couriers. Exploding information—how to process and assimilate it—will be the new headache. "Data bases" will become the communication-age libraries, and researchers will need a computer to poke around in their nooks and crannies.

Likewise, the inventors of the twenty-first century will be more abstract thinkers than their predecessors. No longer mechanics-tinkerers in the mold of Alexander Graham Bell or Thomas Edison, the new-age inventors will be fully versed in the theoretical laws governing the physical universe.

Daniel Bell singles out chemistry as the first truly "modern" technology because chemists must have a thorough knowledge of the abstract properties of macromolecules to manipulate them and hypothesize about what is happening. Indeed, all the science-based industries and products of today—lasers, holograms, electronic equipment, recombinant DNA—wouldn't be possible without years of research and development by highly educated teams of scientists and engineers. And the products of tomorrow will require increasingly more research and development.

"The crucial point about postindustrial society," writes Daniel

Bell, "is that knowledge and information become the strategic and transforming resources of the society, just as capital and labor were the strategic and transforming resources of industrial society. The crucial 'variable' for any such society, therefore, is the strength of its basic research and science and technological resources—in its universities, in its research laboratories, and in its capacity for scientific and technological development."

If the transforming forces of the postindustrial state are knowledge and information, then government policy-makers during this turbulent transition period have to concern themselves with two pivotal questions:

What policies will encourage citizens to acquire the knowledge and information necessary to move the country forward economically and socially? And second, what policies will motivate the country's scientists, engineers, and entrepreneurs to scale the technological heights?

Policy-makers should probably devote the bulk of their resources to solving the latter problem, since the former poses less of a dilemma. Why? Because the very existence of a new technology tends to force people to change their outlook. Certainly this has been the American experience. Take the introduction of television in 1939 and the proliferation of computers in large corporations in the 1960s. No one can dispute the fact that both these events had a profound effect on Americans' expectations, opinions, and life-styles. Thus, progress in the material sense of the word seems to depend upon providing our entrepreneurs—a good portion of whom are scientists and engineers—with a powerful incentive for innovation.

In a capitalist country such as the United States, incentive comes through the promise of individual monetary reward: If an entrepreneur offers a practical new product or service to the public that the public is ready to accept—or can be made to accept through advertising—then that entrepreneur stands a good chance of getting very rich off his or her idea.

The prime mover in the capitalist drama has always been the individual entrepreneur. And to the traditional American way of thinking, an entrepreneur *deserves* to get rich from an extraordinary new product or service because he or she is a person with the foresight not only to breathe life into a vision but also a person with a rare, three-pronged ability: the ability to organize labor, capital, and raw materials efficiently; the ability to marshal information and knowledge and deploy it effectively; and the ability to conquer the brutal forces of the marketplace. Anyone with that triple-threat knack for business surely deserves some reward. And the best capitalism can bestow is glorious piles of mammon.

In the United States, government policies that help individual Americans get rich through entrepreneurship are policies that encourage technological innovation. After all, throughout American history, our notion of success—the so-called American dream—has been to stockpile a personal fortune.

A multitude of theories has been formulated to justify *why* Americans should want to get rich and *how* they should set about doing it. These theories are instructive, for no one can truly say he or she understands the American character—or the American business mentality—until that person understands the myths that drive us onward.

Contradictory as it may seem, the historic American motivation for fortune-building is spiritual. This crusading Protestant ethic, which landed with the Puritans back in 1630, continues to permeate our society to this day.

The Puritans' Calvinist doctrine made pious money-grubbing into a form of salvation. The Puritans believed in a grand predestination plan that held that God had already chosen His Elect—those who would sit by His right hand. Although mortals couldn't influence His decision, a person could prove to his neighbors that he was a member of the Elect by working hard and prospering on this earth. Thus, an individual's material possessions became a sure sign of his eventual welcome into the kingdom of heaven.

A century later, Benjamin Franklin, a printer by trade, seized the Puritan gauntlet and extolled the virtues of thrift, industry, and temperance in secular terms in *Poor Richard's Almanack,* his *Autobiography,* and other popular works of the day.

But the Puritan work ethic reached its fullest flowering during the late-nineteenth-century industrial revolution when Horatio Alger, Jr. (1832–99), the son of a debt-ridden New England parson and a Unitarian minister himself, launched a full-scale propaganda campaign in over a hundred of his best-selling novels for boys. (Gossip to the contrary, Horatio Alger was *not* a Victorian playboy who dissipated his energies and sullied his reputation chasing the female low life through the streets of Montmartre.)

As a motivator of the American masses, Alger has no equal even today. Alger gave his streetwise adolescent heroes memorable names such as Ragged Dick, Phil the Fiddler, and Nelson the Newsboy and endowed them with sterling qualities. They were bright and ambitious yet humble enough to accept their station in life, which always improved steadily throughout the course of their adventures. Naturally, they were scrupulously honest, compulsively frugal, and considerate and helpful to others—especially rich old ladies. They were also annoyingly cheerful in the face of disaster. And every one of them seemed cursed by an overdeveloped sense of fair play. In fact, between the Civil War and World War I, Horatio Alger was probably responsible for more teenaged boys standing up to the neighborhood bully and getting bloody noses and living in mortal fear of foreclosed mortgages than any other human being before or since.

Ironically, not one Horatio Alger hero became a millionaire despite the fact that his sole goal in life was to stay several giant steps ahead of creditors and other villains and eventually to become a creditor himself. Reading these books today, one can't help wondering what Paul the Peddler would do if he had to contend with street peddler licensing fees. And once he made it to small-businessman status, what would he do about business zoning restrictions; minimum wage and hour legislation; workmen's compensation; income tax; Social Security, medical, and other insurance deductions; OSHA regulations; and antipollution statutes? Clearly, times have changed.

As the sociologist C. Wright Mills pointed out so compellingly in his books, the rags-to-riches route into the financial stratosphere always was something of a myth in America. In his seminal work *The Power Elite,* he analyzed the backgrounds of the 275 individuals considered to be the richest people living in the United States over a ninety-year period from the Civil War to 1950. "Poverty-stricken" and "uneducated" are not adjectives that accurately characterize most of their origins.

According to Mills, there were ninety multimillionaires in the group that reached full maturity in the 1890s. About 13 percent of them were foreign-born. Thirty-nine percent were the sons of lower-class people, with another 39 percent coming from the upper class. The remainder came from stable, middle-class homes.

Of the ninety-five superwealthy individuals who reached their maturity in around 1925, only 12 percent had their origins in penury. On the other hand, the upper class had contributed 56 percent, with the remainder from the middle class.

By 1950, a mere 9 percent of the ninety superrich were completely self-made; 68 percent were aristocrats by birth, with the rest from the middle layers of American society.

But it is the educational attainment of these 275 plutocrats that may be the most startling factor of all: Even in 1900, 31 percent had graduated from college; by 1925, 57 percent held college degrees; and by 1950, a full 68 percent were college graduates.

Horatio Alger notwithstanding, it has always been a long shot when a functional illiterate from the boondocks moves to the big city packing nothing more than good intentions and a willingness to work hard and makes it into the financial big time. It will be almost impossible for such a person to prevail against the complex urban and technological forces gripping the United States in the 1980s and 1990s.

Unfortunately, even today, too many pop sociologists and economic observers persist in perpetuating the outmoded notion that the main component of the entrepreneurial success formula is psychological: Do you possess the right mix of personality traits that will propel you across that great financial finish line in the sky?

Sure, it's part of the story. But the question they overlook is even more important: Do you have enough formal knowledge in one of the core disciplines that are responsible for moving our society into the new age?

In a previous book, *The Very Rich Book,* I summarized the list of traits that various studies have identified as those predisposing a person to singular business achievement. They are: excessive energy, the manic kind that keeps you working—or playing—around the clock . . . overweening ambition . . . an addiction to winning . . . an ability to take enormous risks without collapsing under the stress . . . supreme self-confidence and the inner resources to ignore the jeers of skeptics . . . a disciplined mind—that is, basic intelligence, common sense, and the ability to concentrate for long periods and devise innovative solutions to complicated problems . . . a sense, or "hunch," about what will work . . . and an extroverted personality that puts you in contact with influential people who can help you in your arduous climb to the top.

Granted, what is needed to get rich over the next two decades is what has always been needed—ingenuity, persistence, some seed money, and a herculean amount of hard work. A working knowledge of semiconductor technology won't hurt, either. Nor will choosing to apply your entrepreneurial skills in an industry that is young and growing rapidly.

There is nothing new about this concept—*that personal fortunes are usually made in infant or emerging industries*—but it's one that has long been overlooked in the how-to-get-rich literature. A quick survey of American business history is all that is required to prove where the major fortunes were made. (See Table 1.)

During the seventeenth and eighteenth centuries—the preindustrial period in America—men amassed fortunes and moved up the social pyramid the most rapidly in the occupations of trader and merchant. What they traded or sold was not as crucial a factor as it would later become, since the young nation was starved for most goods, particularly manufactured products and scarce commodities.

John Jacob Astor is an excellent example of fortune-building eighteenth-century style, for he is one of our country's earliest success stories. Born in 1763 in Waldorf, Germany, he arrived in Baltimore in 1784 almost penniless. By 1800, he was prospering as a musical instrument peddler and fur trader headquartered in New York City. But the real bonanza would come to him through his American Fur Company, which he chartered in 1808. By 1820, the company exercised a virtual monopoly over that trade in the U.S. territories. When Astor died in 1848, he was one of the principal landowners in New York and one of the wealthiest men in the country. His descendants are, to this day, prominent members of the monied old guard.

The second half of the nineteenth century in expansion-minded America provided just as many, if not more, opportunities for amassing a personal fortune. Anyone who was free, white, and not even twenty-one could find plenty of work, since there were far more jobs available than people to fill them, a condition that persisted until the waves of southern and eastern European immigrants reached these shores at the turn of the century.

TABLE 1 Sources of Personal Fortunes in U.S. 1607–2000

PREINDUSTRIAL AMERICA
INDUSTRIAL AMERICA
POSTINDUSTRIAL AMERICA

1607

The early colonial period is characterized by a feudal economic system under which European monarchs and trading companies awarded huge tracts of land to a handful of aristocrats. These men ruled over their "fiefdoms" as gentlemen farmers/landlords. Some 95 percent of the remaining colonists were also engaged in agriculture-related pursuits either as indentured servants, small farmers, or artisans.

Because there was little specie or paper money in circulation, colonists bartered goods and services. A man's wealth was measured by the size of his landholdings. Using that standard, some of the richest men of the period were:

Sir Ferdinando Gorges (Maine)
Capt. John Mason (New Hampshire)
Van Rensselaers, Schuylers, Beekmans, Van Cortlandts, Livingstons, Morrises, and Philipses (New York)
John Lord Berkeley and Sir George Carteret (New Jersey)
William Penn (Pennsylvania)
Cecilius Calvert, Lord Baltimore (Maryland)

1700

In the eighteenth century under mercantilism, trade and commerce were the quickest roads to personal fortune for the yeomen of the northern colonies. But once acquired, wealth was usually converted into land, still the primary source of political and economic power. Among the richest merchants were:

Andrew Faneuil and Thomas Hancock (Boston)
Moses Brown (Providence)
Pierre Lorillard (New York City)
Richard Bache (Philadelphia)

In the southern colonies, the wealthiest men owned large plantations, tobacco being the principal cash crop. George Washington was one of the richest planters and left a fortune valued at $530,000 when he died in 1799.

William Byrd II, an heir, was perhaps the epitome of the southern tobacco squirearchy. He managed his 180,000 acres from a lordly Georgian mansion, "Westover," in tidewater Virginia.

1789

The new nation threw off the yoke of mercantilism, a system that forbade manufacturing in the colonies. Gradually, Americans' mechanical ingenuity was replacing animal and human *muscle power* and the energy generated by the sun, wind, and water with *machine power* and the energy released by fossil fuels. Unfortunately, it was the rare inventor who got rich. Rather, it was the man—often the merchant with money to invest—who combined an appreciation for things mechanical with business acumen about their application. Typical of this breed were:

Francis Cabot Lowell; Amos and Abbott Lawrence (textiles)
Peter Cooper (iron, telegraph)
John Jacob Astor (furs, banking, real estate)
August Belmont, Francis Martin Drexel, Wm. Wilson Corcoran, George Peabody, Stephen Girard, Ezra Cornell (finance)

Railroads made others rich:
Commodore Cornelius Vanderbilt
John Murray and Robt. Bennet Forbes

1860

The fortunes made during this bustling industrializing phase clustered around core industries:

Steel
Andrew Carnegie
Henry Phipps
Henry C. Frick
Charles Schwab

Mining, Smelting
George Hearst
Meyer Guggenheim
John W. Mackay
James G. Fair
James C. Flood
Wm. S. O'Brien
Tom Walsh
Adolph Lewisohn

Oil
John D. Rockefeller
William Rockefeller
Stephen V. Harkness
Henry M. Flagler
Oliver B. Jennings
Charles Pratt
Henry H. Rogers
Oliver H. Payne
John D. Archbold
Samuel Andrews
Jabez Bostwick
A. C. Bedford

Henry Havemeyer
H. J. Heinz

Railroads
James J. Hill
Jay Gould
E. H. Harriman
Leland Stanford
Mark Hopkins
Charles Crocker
Collis Huntington

Mass Transit
William C. Whitney
P. A. B. Widener
Thomas Fortune Ryan
William Elkins

Machinery and Equipment
John Deere
Cyrus H. McCormick
William Deering
John W. Gates
Isaac Merrit Singer
Edward Clark

Retailing/Real Estate
Potter Palmer
Marshall Field
Alexander T. Stewart

Food Processing
Philip D. Armour
Gustavus Swift

Tobacco
R. J. Reynolds
James B. Duke

1900

By 1900, the United States was producing one third of the world's coal, iron, and steel and led Germany in machine-tool output. Farming, which still remained the single largest segment of the economy, continued to decline.

The era of railroad supremacy was over. Too many railroads had been built, resulting in rate wars and bankruptcies. The financial titans of the early twentieth century—J. Pierpont Morgan, Andrew Mellon, Jacob H. Schiff, George F. Baker, Edward T. Stotesbury, and James Stillman—moved to consolidate and systematize America's recent industrial gains.

The era of the auto had begun with Henry Ford, maker of the ubiquitous Tin Lizzy, reigning as richest man (over $1 billion). Other self-made multimillionaires included:

Autos/Rubber
Horace Dodge
Harvey Firestone
Walter Chrysler

Utilities
Thomas Edison
Nicholas Brady

Food/Beverages
John T. Dorrance
C. W. Post
Kellogg brothers

1929

In fateful 1929, the thirty-eight wealthiest Americans had an average yearly income of $9.5 million each. After the Crash, their annual income was reduced by 25 to 50 percent. But fortunes continued to be made throughout the Depression, even on Wall Street (Joseph P. Kennedy, Alan P. Kirby, Bernard Baruch, Clarence Dillon).

Oil was still gushing new wealth (Pews, Blausteins, William Keck, James Abercrombie, John Mabee, James Chapman, Erle P. Halliburton, Robert Kleberg, Jr.). And during World War II, heavy industry and its captains benefited (Henry J. Kaiser, Cyrus Eaton, Stephen Bechtel, George and Herman Brown, Arthur Vining Davis, Roy Arthur Hunt).

General Motors generated other fortunes (Charles Kettering, Alfred P. Sloan, Charles Mott, John L. Pratt), as did consumer products (Gerard Lambert, Robert Wood Johnson, Eli Lilly, E. Claiborne Robins).

1950

The United States flourished in the first two decades after the war. New multimillionaires were minted by a cornucopia of industries, some never envisioned before:

Aerospace
Juan T. Trippe
C. V. Whitney
Howard Hughes
J. S. McDonnell
Olive A. Beech
William Blakley
Edward J. Daly

Real Estate/Mortgage Lending
William J. Levitt
Peter Kiewit
Eli Broad
Harvey Meyerhoff
S. Mark Taper
Howard Ahmanson
Thomas E. Leavey

Transportation
Daniel Ludwig
Malcolm McLean
Galen Roush
Lykes family

Chemicals/Glass
du Ponts
Houghtons
Olin brothers
Edward B. Osborn
Daniel J. Terra

High Technology
Sherman Fairchild
Chester Carlson
Edwin Land
David Packard
William R. Hewlett
Edwin Whitehead
H. Ross Perot
Cox family
Watson family

Consumer Products
Joyce C. Hall
Leonard Stern
Henry S. McNeil
Ewing Kauffman
Helena Rubinstein
Charles Revson
McConnells/Clarks
Shapiro family

Broadcasting/Publishing
William S. Paley
Samuel I. Newhouse
William Benton
Katherine Graham
Sulzbergers
Annenbergs
Wallaces

1973

The year 1973 will be remembered as the time the oil from the Arabs stopped flowing, at least for a while. American fortunes were still being made in the industry (H. L. Hunt, Sid Richardson, Clint Murchison, J. Paul Getty, Algur Meadows, Leon Hess, Armand Hammer, Marvin Davis, Mitchell family), but the crisis alerted policy-makers to the fact that the supply of this fossil fuel was limited. Other energy sources would have to be developed.

An emerging "third wave" mentality gains credence. People begin accepting the ecological premise that nature will avenge itself if exploited. Therefore, environmental costs must be factored into the commercial equation. A corollary idea is that technology won't necessarily create a brighter, cleaner future unless it's the right kind of technology, preferably electronic-based with its miserly energy requirements and lack of noxious emissions.

Who will be our richest citizens in the year 2000? See the last chapter of this book.

2000

During this industrializing phase, work in most job functions and industries afforded the means to live reasonably well and move up at least one level within the social pyramid. But hard work and ingenuity applied in one of the *newer* industries of the era, one of the adolescent industries that were pushing the country forward into the machine age, promised much more. In those industries, a person could—with any gumption and luck—get rich, maybe even *very* rich. There were plenty of role models, to be sure. There was Staten Island steamboat captain-turned-railroad magnate Cornelius Vanderbilt . . . backwoods, Baptist-bred oil billionaire John D. Rockefeller . . . Scottish immigrant steel buccaneer Andrew Carnegie . . . and later, automobile mass producer Henry Ford.

It's true that these men—none of whom came from illustrious or socially connected families—were clever, ruthless, opportunistic, driven, penny-pinching, and all the other overworked adjectives that are mustered to describe our "robber barons." But they were also *there.* They were in the right place at the right time with the right idea. *Being there* is half the battle.

It's always the right time to get rich in the United States. There never has been a wrong time, even during the long, smoldering Depression years. While the average person merely scratched out a subsistence living, if that, during those years, other people prospered *because* of the hard times. And a few got very rich indeed. Such is the "miracle" of capitalism.

It is the thesis of this book that, given a full quotient of those entrepreneurial traits I outlined earlier and the appropriate educational background—whether acquired through formal schooling or through heavy, extracurricular study—a person in one of the emerging industries of his or her era has a better chance of making a small fortune than his or her entrepreneurial counterparts in mature or declining industries. This circumstance was true in the past and it will be true over the next two decades as the United States moves into the postindustrial age.

Who would you put your money on in the 1980s—an entrepreneur starting up a steel company, or an entrepreneur with a design for a microchip that is tinier and faster than those already on the market? An entrepreneur with well-thought-out plans for manufacturing a new drug using traditional methods, or an entrepreneur with plans to produce the same drug using experimental gene-splicing techniques? An entrepreneur who wants to revitalize a rotting, rusty old railroad line, or an entrepreneur who wants to build a robot with sight and hearing capabilities?

For *The Very Rich Book,* I compiled a list of the 345 richest Americans alive at the start of 1981. These people were each worth a minimum of $50 million, with a few worth over $1 billion.

An analysis of the industries from which their fortunes were generated reveals that oil, real estate/construction, and finance (i.e., insurance; investment and commercial banking; stock brokerage; and private investing) head the list. The other industries represented on the list, and the percentage of the very rich with a stake in them, appear in Table 2.

It's significant that a majority of the 1981 list of the richest Americans were sixty years or older; a handful were in their nineties. Moreover, 43 percent were inheritors. Thus this list still reflects industries that peaked fifty years ago.

Oil tops that list. That industry has been a prime source of personal fortunes ever since the rush on Titusville, Pennsylvania, in 1859. But over the next two decades, oil will lose its hegemony, mainly because there is a finite amount of it left to discover. Other energy sources will gradually replace oil, and then they will begin generating personal fortunes.

By 1981, computer hardware manufacturing had contributed to the personal wealth of only 5 percent of the richest Americans; computer software design and servicing to only 1 percent. Here are two industries that are exploding and will be near the top of the list in the year 2000.

What other industries will be shuffled up or back on the richest Americans list in the year 2000?

That is the question I put to 225 economic and social forecasters in an extensive survey I conducted for this book. (See the Appendix, which reproduces the questionnaire. The respondents' collective opinion is discussed in the Introduction, "Avenues to Wealth by the Year 2000.")

The purpose of the poll was to enable me to predict which of today's entrepreneurs are the likeliest candidates for the richest Americans list in the year 2000. My list of those entrepreneurs appears at the end of this book (see Part VII).

To compile the list, I identified the emerging industries that will

TABLE 2 Industries Represented on 1981 Richest Americans List*

Industry	*Percent of Richest Americans with Substantial Holdings in the Industry†*
Oil/energy	23%
Real estate/construction	20%
Finance	19%
Broadcasting/publishing	12%
Capital goods/heavy industry	6%
Transportation (including airlines and aerospace)	6%
Consumer products (excluding drugs)	6%
Agriculture/ranching	5%
Computer hardware	5%
Food/beverage processing and marketing	5%
Chemicals	5%
Television/film production	4%
Retailing/merchandising	4%
Hotels and gaming	3%
Commodities, minerals/mining	3%
Fast-food and supermarket chains	3%
Forest products	3%
Drug manufacturing and retailing	2%
Computer software	1%
Tobacco	1%
Glass manufacturing	1%
Service (i.e., advertising, research, training)	1%
Fertilizer and seeds	.75%
Trading stamps	.5%
Medical supplies/equipment	.5%
Clothing manufacture	.5%
Textiles	.5%
Biotechnology	.5%
Telecommunications	.15%

*This list contains the names of 345 individuals—and in a few cases, families—worth a minimum of $50 million in 1981. The list was compiled for *The Very Rich Book* by Jacqueline Thompson (New York: William Morrow & Company, 1981).

†When added, the percentages on this table exceed 100 percent because the financial interests of many wealthy people span several industries, often totally unrelated.

sustain the most growth over the next two decades; then I identified the companies that will probably be the industry leaders. The entrepreneurs who founded or own a controlling interest in those companies—and the major investors in those companies—comprise the core of my list of America's next generation of supermillionaires. It is a list, I might emphasize, in which Ph.D.s appear in disproportionate numbers to the percentage that actually exists in American society.

Finally, I included on the list some entrepreneurs engaged in activities outside the predictable high-growth streams but who in my estimation stand a good chance of beating the odds and zooming to the apex of the monetary pyramid anyway, a place where the space is always extremely limited.

ALBRIGHT COLLEGE LIBRARY 196791

Contents

Preface 7

Introduction Avenues to Wealth by the Year 2000 27

PART ONE ***Medical Science Millennium*** 51

1 Biotechnology: **The Breakthrough Wizards** 53

PART TWO ***The Computer Renaissance*** 75

2 Hardware Horizons: **The Computer Manufacturers** 77

3 Computer Parts and Peripherals: **The Suppliers** 106

4 Software: **Its Publishers and Servicers** 128

PART THREE ***Telephones for Tomorrow*** 157

5 Conveying a Zillion Conversations: **Ma Bell and Competitors** 159

6 Switching On: **The Telequipment Tycoons** 178

PART FOUR ***The Stay-at-Home Society*** 191

7 Moving Pictures: **The Electronic Entertainers** 193

PART FIVE ***The Information-Age Office*** 219

8 Crunching Numbers and Processing Words: **The Work-Station Visionaries** 221

9 From Gutenberg to Galactica: **The Space-Age Communicators** 234

PART SIX ***The Factory of the Future*** **249**

10 Blue-Collar Automation: **The Roboticians** 251

PART SEVEN ***Entrepreneurs Who Are Candidates for the Richest Americans List in the Year 2000*** **273**

Appendix Survey Questionnaire 337

Acknowledgments 343

Bibliography 347

Notes 353

Index 376

Introduction: *Avenues to Wealth by the Year 2000*

If two events ever signaled the passing of the old order, they were the epic bankruptcy of the Penn Central Railroad in 1970 and the razing of Henry Ford, Sr.'s, Highland Park factory in 1981.

The Penn Central Railroad was created by a 1968 merger between Commodore Vanderbilt's old New York Central Line and the Pennsylvania Railroad, two of the leading roads in that once-supreme, now-doddering industry. Penn Central was the nation's sixth largest company, and its bankruptcy at that time was the greatest in the history of the United States. The Ford Motor Company's Highland Park factory was the birthplace of the moving assembly line and other innovative mass-production methods, and, as such, could be viewed as the spiritual cradle of the American automobile industry.

Railroads were to the nineteenth century what automobiles are to the twentieth century and what satellites and space shuttles will be to the twenty-first century. But railroads and automobile manufacturing are no longer catalytic economic forces. Replacing them is a new cluster of technologies that are spawning new industries and minting new multimillionaires with the same dynamism as their nineteenth- and early-twentieth-century predecessors.

The ideal time for crystal-ball gazing is when the past and the future seem to bump up against each other, when one era is sputtering to a close and another is just beginning. Economists call it a "period of economic discontinuity." Such a time is now.

Macroeconomists fond of wave metaphors would explain the current transitional turmoil in the American—indeed, the world—economy as a time when one economic wave is receding while another gathers strength. Economists at the Massachusetts Institute of Technology have even constructed an econometric model, the System Dynamics National Model, to explain this basic alternating pattern of social and economic change. The model illustrates the "long wave" economic theories of an obscure Russian economist and unorthodox Communist named Nikolai Dimitriyevich Kondratieff.

Kondratieff was a ranking scholar at Moscow's Agricultural Academy in the early part of this century before the Bolshevik Revolution; he was head of the Business Research Institute after the Revolution. But his ultimate destination was one of Stalin's prison camps in Siberia. There Kondratieff ended his days a heretic for propounding the "reactionary" notion that capitalist depressions are self-correcting rather than terminal.

Kondratieff arrived at his conclusion by studying prices, trade, and production levels in various Western European countries and the United States over a 130-year period. He noted that price indices appeared to rise precipitously and fall just as sharply with a remarkable degree of regularity, which led him to propose what has become known as the Kondratieff cycle or "long wave" theory of economic growth and stagnation.

According to this hypothesis, a long wave generally covers a span of fifty to sixty years and is immediately followed by yet another wave of similar duration. Each wave includes twenty to thirty years of rapid economic expansion and prosperity caused by the flowering of new technologies into rapidly growing industries. When each wave crests, it ushers in another contrasting twenty- to thirty-year period of economic contraction and adversity. But such periods of stagnation are not what they seem, for, ironically, it is during these periods of apparent languor when startling scientific discoveries are made that provide the technological impetus to produce the next wave. (See Table 3.)

The reasons such downturns welcome technological innovation and entrepreneurial enterprise are central to the thesis of this book.

During sluggish periods, the entrenched economic order is weakened. This lassitude invites challenge from progressive, usually younger, more energetic interests. Typically, the economy has experienced some years of zero or negative net investment; factories and equipment are crumbling and outdated; and unemployment is

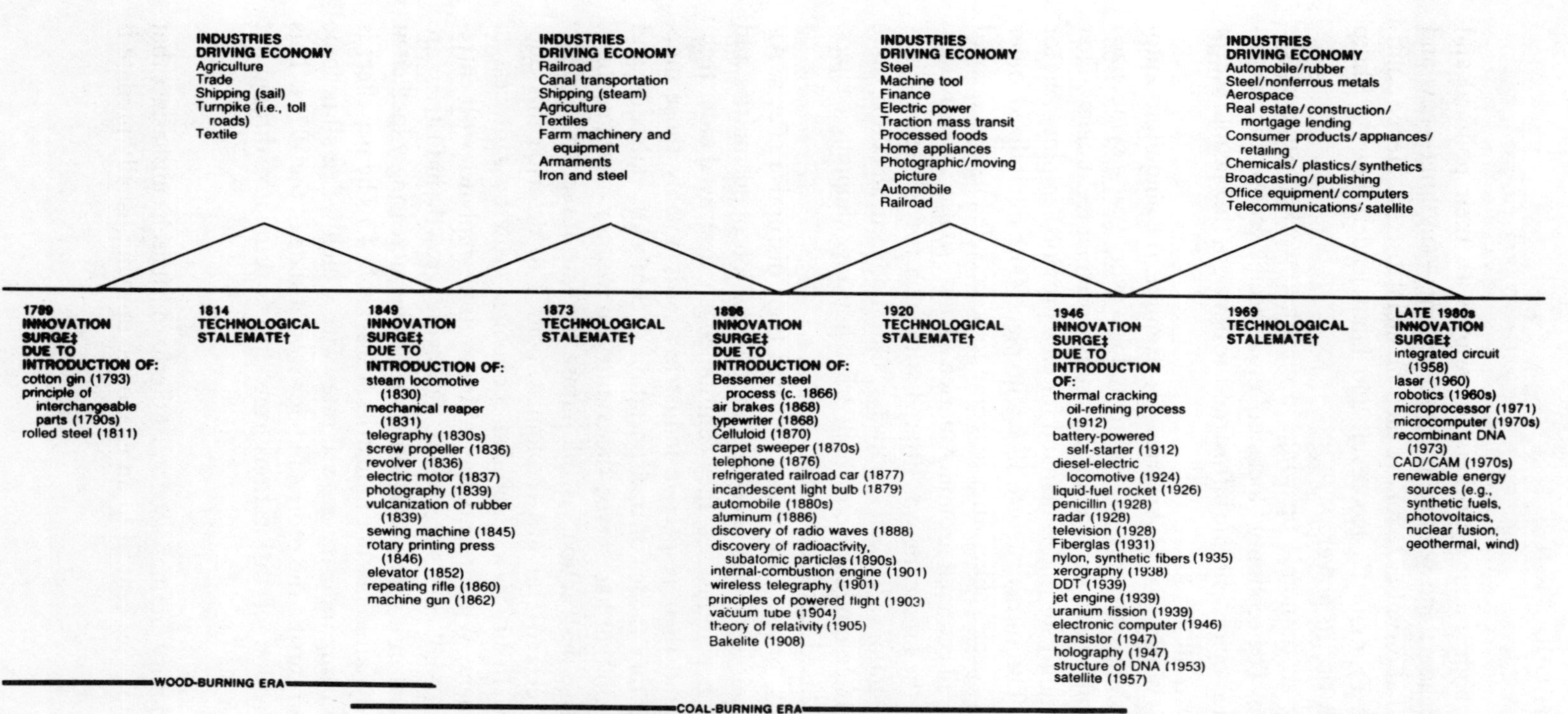

TABLE 3 Long-Wave Periods* of Economic Growth and Stagnation

*The "long wave" theory is often called the Kondratieff cycle after one of its first proponents, Russian Marxist economist Nikolai D. Kondratieff.

†"Technological stalemate" refers to the period between an old technology's maturation and a new technology's takeoff. The concept is ascribed to contemporary German economist Gerhard Mensch, author of *Stalemate in Technology: Innovations Overcome the Depression*. Cambridge, Mass.: Ballinger, 1979.

‡"Innovation surge" is another concept of Mensch, who postulates that during the downturn of a long wave, economic conditions are uniquely propitious for the implementation of new technologies.

painfully high, making workers unhappy with their political and business leaders and willing to try something—anything—new and different that promises to relieve their financial distress. Hence politicians like Franklin D. Roosevelt and labor leaders like A. Philip Randolph rise to power.

During an economic upswing, in contrast, workers support the status quo. The economy "locks in" on a specific group of technologies. And a socioeconomic infrastructure develops that people accept as "the way things are."

In post-World War II America, for example, the ubiquitous automobile—shiny, streamlined, and remodeled every year—symbolized the good life as well as being the economy's premier technology. That era's seemingly insatiable demand for cars provided employment for millions. The infrastructure it set in place both revitalized older industries (steel, rubber/tire, machine tool, petroleum) and created new ones, all centered around America's latest frontier: the suburbs. Suddenly the landscape was littered with such newfangled oddities as filling stations, six-lane highways, automobile dealerships, state motor vehicle bureaus, jerry-built tract housing, shopping centers, and franchised fast-food outlets.

But cars, buses, and trucks also had their disruptive effects. Although their very existence opened up new job opportunities and generated new industries, they simultaneously destroyed older technologies and their respective industries, most notably those nineteenth-century modes of transportation (steamboats, railroads, and trolley cars) that had served the country so well for over a century.

"Creative destruction" is the phrase the late Joseph A. Schumpeter, a revered Harvard economist, coined to describe this phenomenon. In his view, capitalist economies grow by a process of displacement. When technological innovation combines with entrepreneurial skill and investment, emerging growth industries are created that inflict pain and hardship on the increasingly inefficient and less dynamic mature industries. (See Table 4.) The net effect is the contrasting industry growth rates with which we've all become so familiar during the protracted downward slide of the 1970s. This chain reaction of dislocations runs like a *leitmotiv* throughout American history.

Contemporary German economist Gerhard Mensch emphasizes that the technologies central to an economic upswing usually predate it by many years, existing as technically feasible ideas long before they're exploited commercially. Auto magnate Henry Ford, for example, was an entrepreneur whose well-calibrated sense of timing in founding his ventures allowed him to take maximum advantage of the upswing of a Kondratieff wave.

Ford launched his first car manufacturing company in 1899, just three years after the long-wave price bottom. Although that company failed, he founded his second enterprise, the one that survives today, in 1903 with $28,000. For decades that company's expansion was completely financed out of profits because Henry Ford was not only a production genius but also something of a business genius as well. Whether by intuition, coincidence, or forethought (forethought being unlikely), he timed his entry into the marketplace to take full advantage of the sweep of rising prices. It's doubtful he would have been nearly as successful without the expanding prosperity that the upswing of the wave gave to the man in the street, his target consumer, and to the federal, state, and municipal authorities who would build the paved roads and other improvements essential to widen the auto's popularity.

The technology necessary to build the gasoline-engineered car, by the way, was extant long before Henry Ford tried to turn it into a money-maker. While the origins of the motorized vehicle are murky, it is generally accepted that Nikolaus Otto built the first commercial four-stroke-cycle gasoline engine in Germany in 1877 and that Charles and Frank Duryea of Chicopee, Massachusetts, got their one-cylinder phaeton carriage to carry them two hundred feet on September 20, 1893. Henry Ford, on the other hand, didn't produce a truly successful car, in the commercial sense, until late 1908, when the fabled Model T rolled out of his plant. By 1926, more than fifteen million of these Tin Lizzies had been sold, some for as little as $290, making old Henry the premier car manufacturer in the world.

The phenomenon of the technological lag suggests that the important scientific breakthroughs inspiring the next long-wave surge of the 1990s have already occurred. In fact, Gerard O'Neill, a Princeton University physics professor as well as the president of the Space Studies Institute and author of *2081,* a nonfiction book describing life in that year, makes a much bolder assertion. He claims that "over the span of a century, the future is likely to be shaped far more by the evolution of technologies we already understand than by the

TABLE 4 Mature Industries Represented by Their Domination of the Top Ten Places on the Fortune 500 List*

1960† Fortune 500 List (Top Ten Corporations by Industry)‡		*1970† Fortune 500 List (Top Ten Corporations by Industry)‡*		*1980† Fortune 500 List (Top Ten Corporations by Industry)‡*		*Hypothetical Fortune 500 List for the Year 2000§ (Top Ten Corporations by Industry)*	
Oil refining	40%	Oil refining	40%	Oil refining	60%	Energy/mining/oil	40%
Automobile	30%	Automobile	30%	Automobile	20%	Computer hardware	30%
Steel	10%	Electronics/appliances	20%	Electronics/appliances	10%	Telecommunications	10%
Food	10%	Office equipment/computers	10%	Office equipment/computers	10%	Electronics/components	10%
Electronics/appliances	10%					Miscellaneous	10%
	100%		100%		100%		100%

*The *Fortune* 500, compiled and published for the first time in 1955 by *Fortune* magazine, lists the top five hundred U.S. industrial corporations ranked by gross sales. *Fortune* publishes separate lists of the fifty largest commercial banks, life insurance companies, diversified financial institutions, retailers, transportation companies, and utilities.

†The annual *Fortune* 500 list reflects corporate sales for the previous year. Thus the 1960 *Fortune* 500 covers revenues for 1959.

‡The telecommunications industry, represented by AT&T, would have captured the top spot on the 1960, 1970, and 1980 *Fortune* 500 lists if the *Fortune* editors included utilities on the list of the country's largest industrial companies instead of ranking them separately. Thus telecommunications would represent a 10 percent share of the top ten *Fortune* 500 corporations.

§The "Hypothetical *Fortune* 500 List for the Year 2000" represents responses to the first question on the questionnaire sent to futurists/forecasters to gather data for this book. (See Appendix.)

effects of scientific breakthroughs that we do not yet even suspect." O'Neill isolates computers, automation, space colonies, energy, and communications as the five "drivers of change" he thinks will determine the course of the next hundred years.

What key technologies do the 225 forecasters who participated in the survey for this book (see the Appendix) cite as those they expect to exert the most influence up to the year 2000? Their answer: genetic engineering, telecommunications, microelectronics, and robotics.

In the case of telecommunications, the basic technology was "discovered" over a century ago. The origins of the others go back to the 1950s. Note that these technologies were forced into their development phase almost immediately, for it is an axiom of our age that the speed with which new discoveries are turned into useful products is accelerating.

On the premise that the fastest-growing industries afford the greatest opportunity for entrepreneurial aspirants to haul in the wealth of Croesus, I asked survey respondents three basic questions:

1. Over the next twenty years, what industries or industry segments will grow the fastest, and in what year do you anticipate they will mature?
2. What companies are currently the leaders in these rapidly growing industries/segments?
3. Which of these companies are "entrepreneurial"—that is, in existence less than ten or twelve years and still headed by their founders?

The collective answer to the first question is outlined in Table 5. Note that the industries—genetic engineering, computer software, automated equipment, telecommunications, computer hardware, energy/mining/oil, and electronics and components—could all be subsumed under the label "high technology" (with the possible exception of the energy sector).

Contemporary experts concur that high technology has become the force driving the economies of developed countries. "It is continually changing the matrix of our society," observes Hans Mark, former secretary of the Air Force and director of the National Aeronautics and Space Administration. "It creates industries and

TABLE 5 Highest-Growth Industries, 1980–2000

	Rapid-Growth Industries (in order of highest growth)	*Estimated % Growth* (1980–2000)*	*Pace of Technological Change*	*Year Industry Will Mature†*
1.	Genetic engineering •drugs and other products manufactured using bioengineering production methods	948% or 12% average annual growth rate compounded	very fast	2026
2.	Computer software •program creators/authors •software publishers •software retailing	762% or 11% average annual growth rate compounded	very fast	2003
3.	Automated equipment •office •factory	640% or 10% average annual growth rate compounded	very fast	2004
4.	Telecommunications •data hardware •data transmission •voice hardware •voice transmission •satellite services •terrestrial services	406% or 7% average annual growth rate compounded	very fast	2004

5. Computer hardware •mainframes •minicomputers •personal computers •peripheral equipment	362% or 7% average annual growth rate compounded	very fast	2000
6. Energy/mining/oil •coal •natural gas •petroleum •mining and drilling equipment	358% or 7% average annual growth rate compounded	moderate	2011
7. Electronics and components •microprocessors •semiconductor memory •measurement/scientific equipment	242% or 5% average annual growth rate compounded	very fast	1998

*An "emerging growth company" is generally defined by its size and growth rate. It is a company with less than $100 million in sales whose profits are growing at a rate of at least 15% average annual compounded. The figures in this column refer to aggregate gross revenues by industry.

†In other words, the year the industry reaches a growth plateau before it begins to decline.

jobs that most people—and that includes economists—can't even imagine."

High technology's thrust comes from its two-pronged ability to spawn progressively more startling innovations—such as artificial intelligence, sophisticated medical diagnostic equipment, and outer-space exploration—and at the same time to help old-line manufacturing companies improve productivity and reduce costs.

What is "high technology"?

Some people refer to it grandly as "Man's best hope for a brighter tomorrow." One high-tech company uses as its slogan "The place where science gets down to business," which is, perhaps, somewhat closer to the mark. The editors at the Smithsonian Institution in Washington describe high technology as "a new sort of technology involving a high degree of scientific training and considerable investment in refined apparatus for its exploitation." They point out that by its very definition, the high technology of today will be the low technology of tomorrow.

Leave it to the U.S. Bureau of Labor Statistics to come up with a quantifiable definition: According to the bureau, of the 977 industries assigned standard industrial codes, 36 qualify as "high tech" because their research and development (R&D) expenditures and number of technical employees run twice as high as the average for all U.S. manufacturing. The bureau labels an additional 56 industries "high-tech-intensive" because their R&D spending and technical employment are above the national average.

Clearly, high technology does not refer to a single industry but a cluster of industries the products and services of which range from data processing and computers to medical equipment, from integrated circuits to satellites, from videodisks to genetically engineered pharmaceuticals. And most high-technology companies are founded by people with at least a working knowledge of science and engineering.

Survey questions two and three, seeking to affix names and faces to our emerging generation of entrepreneurs, presuppose, of course, that an entrepreneurial renaissance is in progress. It is. Proof of such a revival is easy enough to muster.

Under both Ronald Reagan and Jimmy Carter, policies designed to help small businesses and foster entrepreneurial activity were enacted. This happened despite the fact that Carter ostensibly supported the "seep up" method of income distribution while Reagan embraces the "trickle down" formula. But in practice both of them seem to agree that, economic theory notwithstanding, our peculiarly American institution—the self-made man—deserves every chance to triumph.

Beginning with the Carter administration, the federal government began giving small business the kind of attention, if not the appropriations, reserved in the 1960s for eradicating poverty and probing outer space. In 1978 and 1980, two fact-finding bodies (the Congressional Subcommittee on Antitrust, Consumers, and Employment, and the White House Conference on Small Business) met to examine the health of small business in America. Meanwhile, Carter appointed the nation's first small-business ombudsman with the title chief counsel for advocacy of the Small Business Administration. The Small Business Administration itself got more funds and soon nearly all of the hundred largest cities in the United States had offices of economic development. These departments are run by a new breed of trained professionals with the mandate to attract new businesses to the area and keep the existing ones from relocating.

The sudden resurgence of interest in America's smaller companies was prompted in part by a worrisome government study showing that companies other than the *Fortune* 1,000 accounted for the vast majority of new jobs created in the country between 1969 and 1976 and that even government civilian employment accounted for more jobs than huge corporations.

Another alarm sounded when officials took stock of the nation's declining venture-capital resources, a prime indicator of entrepreneurial well-being. In 1969, Congress increased the maximum tax on long-term capital gains from 25 percent to 49 percent. That year venture-capital funds stood at $171 million. By 1975, the figure had dropped to a meager $10 million, and it didn't start rising again significantly until Congress rolled back the capital-gains tax rate to 28 percent in 1978. Later, President Reagan rallied the necessary support to lower the rate even further, to 20 percent. Concurrently, the Securities and Exchange Commission liberalized its rules concerning the sale of unregistered stock and simplified the registration procedure for companies seeking to raise relatively small sums of public money.

Thus, by 1983, $2 billion in venture funds was available, more than two hundred times the amount of only eight years earlier. The replenishment of this investment store came none too soon, since the soaring interest rates of the period ruled out the use of bank loans and bond issues to underwrite new businesses.

But if the traditional sources of debt financing were off-limits to struggling entrepreneurs, the equity markets couldn't have been more enthusiastic. The raging new-issues markets of 1980, 1981, and 1983 created a number of instant multimillionaires (Steven Jobs, Steve Wozniak, and Michael Markkula of Apple Computer; Robert Swanson and Herbert Boyer of Genentech, to name a few). In fact, the venture-capital and stock-market thirst for equity in as yet unproven high-technology companies was so strong that many founders began taking them public before their enterprise had even marketed a product and long before it showed any hope of turning a profit.

The popularity of such extremely high-risk ventures was no doubt fed by the media that entered the decade bemoaning the lagging productivity of the country's factories and offices while simultaneously hawking the virtues of America's transcendent gospel, the free-enterprise system, and its traditional savior, the self-reliant entrepreneur. Four slick consumer magazines extolling entrepreneurial risk-taking made their debut around this time in a further effort to raise the public consciousness.

Indeed, the frenzy in the equity markets in the early 1980s reached such a fever pitch that, at one point, a joking commentator suggested a new sure-fire way to get rich: "Save all the hard work that starting a new business entails," advised Jerry Pearlman, a Zenith Radio executive, "and simply change the name of your corporation. The ideal name to change it to is International Genetic CAD/CAM Computer and Cable Television Unlimited."

When Ronald Reagan moved into the White House in January 1981, one of his first decisions was as portentous as it was trivial. He ordered the portrait hanging in the Blue Room of the United States' third President, Thomas Jefferson, replaced by one of Calvin Coolidge, the latter-day President whose most memorable utterance was "The business of America is business." The gesture was a bellwether of things to come. Henceforth, Ronald Reagan and his conservative allies in Congress did everything in their power to turn back the clock and return the country to the decentralized state it was in under Silent Cal.

Many of the Reagan administration's policies were welcomed as a boon for business, particularly big business. A conservative Congress helped Reagan reduce corporate tax rates, accelerate the depreciation schedule for new capital equipment, and allow companies to sell their investment tax credits. It cut social programs and diverted federal funds to the Pentagon, funds that ultimately found their way to defense contractors, large and small. It lifted the federal regulatory "burden" on a host of industries. And Reagan let it be known to the antitrust division of the Justice Department that he did not equate "bigness" with "badness," paving the way for the settlement of long-pending monopoly suits against IBM and AT&T, two of the world's largest corporations.

The Reagan administration also wielded its budget-cutting ax in the direction of federal government support for basic research. (Over the past three decades, government R&D spending has accounted for 70 percent of the nation's basic research.) The Reagan forces did it by narrowing the definition of "fundamental scientific research." In the administration's view, fundamental scientific research seldom includes any activities aimed at commercializing discoveries, except when "long-term, high-risk research is likely to offer broad societal benefits with little commercial payoff." Using the "broad societal benefits" yardstick, a promising alternate energy source, nuclear fusion, is one of the few technologies that might qualify for government support.

Fortunately, American companies—particularly the research-intensive high-technology firms—had already increased their R&D outlay on their own initiative in the late 1970s. This change of direction occurred none too soon, for industrial R&D as a percentage of GNP had been steadily declining since 1968.

To encourage this fortuitous investment trend, Reagan backed legislation giving companies a 25 percent tax credit for increases in their R&D expenditures. The administration's policies also turned R&D limited partnerships into a wondrous, low-risk financing tool for high-technology companies. These R&D tax shelters allow high-tax-bracket investors to underwrite corporate R&D and take whopping tax deductions while holding out the promise of big returns to them should viable products eventually be developed and sold. Trilogy Systems Corp., a 1980 Silicon Valley start-up, is one example of a computer firm that raised $55 million of its $160 million initial capitalization from an R&D limited partnership.

The implications of the Reagan policies were obvious: Sink or swim. Reagan did not intend to prop up weak companies, industries, and regions of the country. He meant rather to unleash the more dynamic forces of the private economy.

Survey question three, seeking to identify specific entrepreneurs whose companies were founded no more than ten or twelve years prior to the publication of this book, is based on another arguable supposition—that the early entrants in emerging industries will still be in the race when their industry matures.

There is an old saying in the business world: Pioneers are the people with the arrows in their backs. How true is it?

In general, fledgling companies selling established products fare poorly when they're competing against large, mature companies. On the other hand, large companies usually succumb to young, entrepreneurial ventures when the product or service in question is highly innovative. In short, *Fortune* 500-size companies, hampered by their sluggish decision-making apparatus, rarely move fast enough to take advantage of the breakneck pace in the high-technology marketplace, at least when a new invention is first commercialized. The IBMs sometimes catch up later, though.

It's important here to distinguish between *discovery* and *innovation,* between scientific breakthroughs (the end result of years of intensive and expensive R&D) and innovative products.

"Breakthrough research is the province of big companies and governmental and university laboratories. The frontiers of science are seldom expanded by the garage-shop entrepreneur," contends Stanford University engineering management professor Henry E. Riggs, an expert on small, high-technology firms. "On the other hand, while Bell Laboratories and others in major research laboratories were responsible for the *invention* of solid-state electronic devices, the *exploitation* of solid-state technology has been largely accomplished by technical entrepreneurial ventures."

Recent history underlines Riggs's observation. The smaller, entrepreneurial companies in the high-technology field have been the source of the most ingenious products. For example, once the transistor, invented at Bell Labs, became available in the 1950s, one might have expected the established computer companies—the Sperry Rands, Honeywells, and IBMs—to act on the obvious equation: smaller=faster=cheaper=better. Instead, they continued to produce oversized, mainframe computers that occupied rooms the size of aircraft hangars, consumed enormous amounts of power, and required controlled environmental conditions—purified air and a cool, even temperature—to operate without glitches. Although RCA introduced the world's first fully transistorized computer in 1959 with the other mainframe manufacturers nipping at its heels, it was the upstart Digital Equipment Corporation (DEC) that pioneered the next radical step in hardware design: the minicomputer.

A thirty-one-year-old, MIT-trained *Wunderkind* named Kenneth H. Olsen founded DEC in 1957 with $70,000 from the granddaddy of venture-capital firms, the Boston-based American Research and Development Corp. He and an associate, Harlan Anderson, set up shop in Maynard, Massachusetts, in a Civil War-vintage woolen mill that produced blankets for the Union Army and where the floors still oozed lanolin on hot days. For the first couple of years, the two of them did most of the dirty work themselves, including cleaning the johns, designing the tools, and knocking on doors selling their initial product—circuit modules for testing computer equipment. These circuit modules became the building blocks for their first "minicomputer," a tag borrowed from the miniskirt craze of the 1960s and not attributable to Olsen or anyone else at DEC.

Not only were DEC's minicomputers smaller, they were also cheaper; their first model sold for $18,000 in 1965 when competing mainframe machines were going for over $100,000. By combining the advantages of transistors and integrated circuits (invented in 1958 by the Fairchild Semiconductor Co.), DEC was able to do something that seemed contradictory: compress a machine's size while simultaneously increasing its computing ability. IBM did not develop a comparable machine until 1976, by which time Data General, Datapoint, Varian, Interdata, and several other novice companies had long since rushed into the fray, and DEC had captured a 37 percent share of the minicomputer market.

DEC survived in the land of the giants using the only tried-and-true formula that seems to work for emerging high-tech firms: It picked a niche that, at least initially, was too small to interest the Goliaths of the industry. DEC's first shipments went to original-equipment manufacturers that packaged computer systems for end users, and to sophisticated scientific and industrial customers for whom price and performance were more important than hand-holding and accompanying software. And DEC fended off competition by constantly improving its products.

Today this two-pronged pioneering formula accounts for DEC's almost $4 billion in annual sales, its No. 84 rank on the 1983 *Fortune* 500 list, and Olsen's personal fortune of close to $250 million.

There are definite advantages accruing to the savvy entrepreneur leading the pack:

A small, innovative company introducing a startlingly new, state-of-the-art product will have the market all to itself for a considerable length of time. The smart entrepreneur uses this interval to consolidate his gains before the upstart companies with their copycat products jump into the marketplace. (Normally it will be several years down the road before an innovator-entrepreneur has to worry about competition from the leviathan companies.) Thus, when the knock-off products do start flooding onto the market, our entrepreneur still has the strategic advantage. He can bank on an established, brand-name product; consumer loyalty; an intimate knowledge of consumers' real needs and desires based on actual experience marketing the product to them; and his hefty market share.

The list of pioneers and early entrants in high-technology fields who have capitalized on such strategic positioning and remained industry leaders is long.

In the aircraft manufacturing field, there are really two vanguard companies. Boeing designed what most people consider the first "modern" airliner, the 247. It entered service in 1933 but was swiftly outdated by the Douglas DC-2. When propeller-driven, piston-engine planes were rendered obsolete by jet engines, it was Boeing's 707 in 1954 and its chief competitor, the Douglas DC-8, that pushed the industry with a whoosh into the jet age. Today Boeing and McDonnell Douglas, the successor firm to Douglas Aircraft, are still front and center in their industry.

Xerography, a dazzling advance in the way offices are run, exemplifies one man's persistent struggle against overwhelming financial odds and the skepticism of the business establishment. Chester Carlson, a patent attorney with a physics degree from the California Institute of Technology, filed for his first patent associated with xerography as early as 1937. But it wasn't until 1947—after Carlson had been turned down by more than twenty major companies, including IBM—that the Haloid Co., a small manufacturer of photographic paper, and a nonprofit research organization put up R&D funds to develop the process. And it was another twelve years before the first Xerox machine, the 914, was introduced to awestruck consumers.

When Carlson died in 1968, he was worth $47 million, and the Xerox Corp. was the undisputed leader in the copier industry. Xerox's market share, once 90 percent, has since plunged to around 40 percent, but it is still the industry leader.

Similar claims can be made about Dr. Edwin Land's self-developing cameras and his Polaroid Corp., which continues to dominate that market. More recently there's the example of physicist Dr. Robert Noyce.

Dr. Noyce began his career in the mid-1950s with Shockley Semiconductor Laboratories, founded as a division of Beckman Instruments by William Shockley, one of the three Nobel Prize-winning inventors of the transistor. But it wasn't long before Noyce and seven of Shockley's other top employees—referred to thereafter as "the traitorous eight" by the embittered Shockley—defected to start Fairchild Semiconductor Co. with backing from the Fairchild Camera and Instrument Co. back East. In 1958, Fairchild and Texas Instruments simultaneously invented the integrated circuit, the product heralding the next leap forward in the computer miniaturization trend.*

With the spin-off phenomenon becoming an accepted entrepreneurial pattern in high-technology circles, Dr. Noyce later left Fairchild to start his own semiconductor firm with Gordon Moore

*Today, Jack Kilby of Texas Instruments, a company that originally produced geophysical equipments for the oil industry and gradually evolved into an electronics firm in the 1950s, is given credit for actually inventing the integrated circuit, while Noyce is known as the scientist who improved it and conceived the planar process for manufacturing it.

and Andrew Grove. Their 1968 start-up venture, Intel Corp., achieved its crucial breakthrough—the 4004 microprocessor chip—three years after its founding. Intel followed that up in 1973 with a speedier eight-bit microprocessor—the 8008—that made typewriter-size computers like the Apple a possibility. Only eight years after its founding, Intel was the world's foremost producer of fingertip-size semiconductor memory devices and could boast responsibility for launching the microelectronics revolution.

Certainly there are equally compelling examples of vanguard companies that have fallen by the wayside. Remember the Bowmar Brain, the toy-size calculator that relegated ten-pound adding machines and slide rules to the garbage dump? The Bowmar Instrument Corp. advertised so heavily that everyone wanted to buy a Brain—and suddenly every established consumer electronics manufacturer wanted to market its own version. The stampede squeezed Bowmar out of the business entirely.

The story was equally tragic when, in the late 1960s, Viatron and Cogar, two virgin computer companies, got the bright idea to turn "dumb" desktop computer terminals into working computers, what we now call "microcomputers" or "personal computers." Viatron attempted to build its own version of a microprocessor, the vital brains for its proposed machine, and managed to spend $30 million to generate $3 million in income, all in the year 1970. Viatron went bust soon after. Cogar almost followed suit but was acquired on the brink of bankruptcy by the business systems division (now sold off) of the Singer sewing machine company.

Docutel (today known as Docutel/Olivetti Corp.), which introduced the first automated bank teller machine in 1971, was another entrepreneurial company that almost died before cashing in on its innovation. Because the tiny Dallas company was slow to incorporate technological advances into its early machines, it quickly lost its market dominance to powerful Johnny-come-lately competitors. By 1977 Docutel was near collapse but has since gotten a new lease on life through a smart product-upgrading move, several savvy financial maneuvers, and an attempt to diversify into other office products. Nevertheless, its slide to a 20 percent share of the automated teller market may be the best it will ever do again.

When innovator companies do fail, the reasons are usually apparent, as they certainly are in the cases cited above. Viatron and Cogar were long on vision and short on cash to underwrite an idea that another pioneering firm, Intel, actually brought to fruition. Bowmar, on the other hand, ignored the niche precept, naïvely believing it could corner a mass market that included every American consumer over age three while the big guys sat back and watched. Docutel simply lost its technological edge.

Surprising as it may be, technically oriented start-ups that are the early entrants in a field have relatively low failure rates compared to other types of new businesses. In fact, during the early 1980s recession, when Dun & Bradstreet's Cassandra list of business failures made the headlines monthly because the rate kept rivaling 1930s Depression levels, the industries most represented on the lists were invariably craft-oriented industries left over from an earlier age—furniture, textile, leather and shoes, and apparel.

Perhaps the best and most up-to-date study of high-technology company attrition was done by Albert V. Bruno and Arnold C. Cooper, two professors of management. The study was conducted over a twenty-year period, and their sample included 250 firms, virtually all of the high-tech companies started on the San Francisco peninsula known as "Silicon Valley" between 1960 and 1969.

In 1980 they did a follow-up survey and discovered that 36.8 percent of the original sample had folded. This is not a high failure rate when, according to Dun & Bradstreet, two out of three small businesses succumb within the first four years of their founding. A total of 30.8 percent of the companies in Bruno and Cooper's study survived as independents, and 32.4 percent were acquired.

The sample firms' acquisition rate is as high as their failure rate is low. The mean age of bought-up companies was 6.4 years, and the buyer company was generally in the same business already, leading to the conclusion that industry shake-outs tend to occur about four to six years after a sunrise industry makes its appearance.

The typical acquired firm was privately held, and all of its founders were present at the time of its acquisition. It's noteworthy that none of the founders interviewed about their company's acquisition admitted to engineering the sale, nor did any suggest that acquisition was part of their long-range strategy. However, Bruno and Cooper inserted a disclaimer:

"Although not represented in these responses, we suspect that owners of successful high-technology firms sometimes chose to 'cash in their chips' by selling out to a publicly owned company. They could thereby become 'instant millionaires,' diversify their portfolios, and open a variety of future options, including becoming entrepreneurs again. Possibly, these responses were not encountered in the survey because such former owners were on their yachts or involved in starting other new firms."

No one would deny that money lust is certainly one of the subterranean rivulets bubbling through the psyches of entrepreneurs in hotbeds of new business activity such as Silicon Valley. Peter Sprague, the wealthy venture capitalist of New England lineage who rescued National Semiconductor from oblivion in the 1960s, tells about the time he tried to arrange a deal for a young California entrepreneur. The man rejected the offer outright the moment he realized he'd make only $3 million on it the first year.

"This guy had about eight dollars to his name," Sprague recalls. "But he had a potentially marketable product and a dream. That's the psychology of Silicon Valley. People don't want to wait for success. They want it now. They want to go straight from the Toyota to the Lear Jet."

Most principals of high-tech firms get one, if not dozens, of opportunities to sell out even if their company is tottering. Actually, a list in the corporate vessel might prove an attraction to a potential suitor with the financial resources of an Allied Corporation, for example. Large, profit-rich companies often welcome a tax-loss carry-forward. Besides, a negative balance sheet should lower an entrepreneurial company's sale price.

Today there are usually several founders, not just one, to consult about a potential sale. Bruno and Cooper describe the typical founding group: Two principals is common, although often there is a group of "early founders" and another group of "late founders," with the latter contributing on a part-time basis until the company can support them full-time. Generally the founders have worked together earlier as employees of an established high-technology company where they became increasingly disenchanted with their jobs. The majority were in their thirties when they decided to strike out on their own for the first time. They are invariably well educated with at least a college degree, often with graduate degrees. Electrical engineering is the discipline that shows up most frequently on their résumés. And a disproportionately high percentage of the founders had fathers who were in business for themselves.

When the founders agree to sell out to a corporate giant such as the one for which they originally worked, the reasons run the gamut. Some entrepreneurs want to use the money they'll earn to start over and, perhaps, build an even better company the second time. Others couldn't stand the intense competition that success engendered and would rather clip coupons and keep in constant communication with their stockbroker from a telephone next to their Olympic-size swimming pool.

Many entrepreneurs are what Columbia Business School professor Avraham Meshulach calls "artisan entrepreneurs." Their expertise is strictly on the technical side; they're managerial lightweights. They can put together mechanical puzzles but they're stumped by financial puzzles. Very often they lose control of their company to venture capitalists as a prelude to the sale. The "managerial entrepreneur," in contrast, is envigorated by the challenge of overseeing a company's rapid expansion. This type is more likely to stay at the helm indefinitely, master of his or her own destiny through fair weather or foul.

A high percentage of the high-tech entrepreneurs who sell out, try again another day. The phenomenon of the entrepreneurial encore is not new. The nineteenth-century business landscape was littered with second- or even third-time ventures of men such as Eli Whitney, Sam Colt, Isaac Merrit Singer, and Andrew Carnegie. Today's high-technology repeaters—Gene Amdahl, Alan Shugart, Ralph Ungermann, and others—exemplify the same basic entrepreneurial pattern. It's an exhausting way to riches, but whoever said the Silicon Valley freeways—or Route 128 near Boston, for that matter—are paved with gold?

Alan Shugart, for one, thinks "It's a waste of talent for a good entrepreneur to start only one company."

He might be right. It's also absolutely essential that an entrepreneur recycle his talents if a place on the year 2000's list of wealthiest Americans is, indeed, his or her goal. One $20 million sale won't do it. In fact, if an entrepreneur does not have what it takes to stick with his company until it breaks the $200 million sales barrier, he might as well resign himself to starting a series of little companies until the small change from selling them off finally adds up to Midas's millions.

Money is always there,
but the pockets change.

—GERTRUDE STEIN

PART ONE

Medical Science Millennium

1 Biotechnology: *The Breakthrough Wizards*

In April 1953, the prestigious scientific journal *Nature* published a seemingly innocuous letter from James D. Watson and Francis H. C. Crick, two biomolecular researchers working in Cambridge, England. The letter briefly outlined their hypothesis concerning the double helix, or spiral, structure of deoxyribonucleic acid (DNA), the molecule that has embedded in it the architecture of life.

Just before they went public with their findings, Dr. Watson had confided their theory to Dr. Max Delbruck of the California Institute of Technology, one of the foremost researchers in the field. Delbruck wrote back:

"I have a feeling that if your structure is true and if its suggestions concerning the nature of replication have any validity at all, then all hell will break loose, and theoretical biology will enter a most tumultuous phase."

Their notion was proven true. And all hell did break loose, for Watson and Crick's discovery set in motion a chain reaction in the field of biology as awesome as the earlier revolution in physics that eventually unlocked the power of the atom. Their hypothesis afforded the provocative clue that sent hundreds of other curious researchers scurrying back to their labs. There they took a fresh look at the genetic nuts and bolts, the chemical cogs and wheels that comprise life on earth. The results of some of that research are nothing short of miraculous.

Just as Drs. Watson and Crick had described, DNA molecules—the chemical substance within each chromosome that carries the organism's hereditary code—are shaped like long, twisted ladders. These DNA ladders do, indeed, direct the production of proteins, the building materials of life. They do it by means of genes. Genes, or segments of DNA, are the individual units of heredity that determine particular characteristics—in the case of a human being, such things as height and hair color. Each gene carries instructions for making a specific protein, especially the proteins known as enzymes. These crucial specks of life are the catalysts that trigger the myriad of chemical reactions within cells.

Throughout the 1960s, there was a succession of breakthroughs in the field of genetics. According to one commentator, collectively they constituted one of the greatest conceptual transitions of all time, comparable to the notion that the sun, rather than the earth, is the center of the solar system.

For example, in 1962 scientists postulated the existence of restriction enzymes that act as scissors, snipping out specific gene sequences from the DNA strands. Four years later, another milestone was passed when researchers deciphered the genetic code or the meaning of particular gene sequences.

But the tumultuous phase Delbruck foresaw didn't take place until after 1973. That was the year when American biochemists Dr. Herbert W. Boyer and Stanley N. Cohen let the world know about the technique they had developed for recombining genes from different organisms—even different species—in bacterial cells. They showed other scientists how to remove chemically a selected gene sequence from one cell and insert it into another. Because the transplanted gene sequence contains all the information necessary to produce a substance organically, the injected cell becomes, in effect, a miniature factory and an incredibly efficient one at that. Under the right conditions, within twenty-four hours that original cell could subdivide into over two hundred trillion bacteria, all busy generating the desired end product.

Boyer and Cohen's technique—referred to variously as "gene-splicing," "genetic engineering," "bioengineering," or "recombinant DNA"—is the biological equivalent of the splitting of the atom. Overnight it transformed a series of scientific advances into a commercially viable technology.

This fact was not lost on Boyer, Cohen, and their respective employers, the University of California at San Francisco and Stanford University. To this day, none of the parties to that discovery have lost sight of the millions to be made by patenting this fundamental process and licensing it to others.

The commercialization of genetic engineering is a paradigm of capitalism at its most avaricious and exemplifies the problematic process by which a basic science becomes an applied science. To be sure, the obstacles blocking the development of bioengineered products are particularly severe:

First, biomolecular research requires a huge reservoir of R&D funding to turn small-scale laboratory production into a large industrial operation. Second, the extensive testing required to win federal Food & Drug Administration (FDA) approval can take up to five years, which may not be worth the effort unless another government agency, the U.S. Patent and Trademark Office, cooperates by granting your product patent protection. That can be an equally drawn-out ordeal.

The rush to profits actually predated Boyer and Cohen's disclosure by two years. It began with the founding of Cetus Corporation in Berkeley, California, in 1971. Although Cetus was the first company whose sole mandate was developing products based on DNA research, it is by no means the most innovative. Cetus's three founders were brave and pioneering, though, for they established their beachhead in this emerging industry at a singularly unpropitious time.

Throughout most of the 1970s, the public still viewed genetic engineering as a harbinger of monsters rather than miracles. There were bills before Congress to regulate genetic research, and local ordinances were proposed to ban it entirely. The scientists working in the field were themselves concerned about possible health hazards should a cloned viral strain accidentally escape from a lab and cause an epidemic of a heretofore unknown disease. As a consequence of this heated public debate, the National Institutes of Health, one of the country's major scientific funding sources, did eventually adopt a set of safety guidelines for biotechnology researchers.

Investors in those early years looked upon bioengineering as a technology with about as much immediate practical application as intergalactic travel. To some extent, they were right. It would be exactly a decade after Boyer and Cohen's 1973 discovery before the first bioengineered commercial product hit the United States market. And it will be several more years yet before even the best companies in the industry begin showing a regular profit.

But none of these obstacles deterred Cetus founders Drs. Peter J. Farley (M.D., St. Louis University; M.B.A., Stanford), Donald A. Glaser (Nobel laureate in physics and molecular biologist), and Ronald E. Cape (Ph.D. in biochemistry, McGill; M.B.A., Harvard) when they joined forces to, in Farley's words, ". . . take biology wherever it will lead us." In the intervening years it has led Cetus through various triumphs and pitfalls that typify the experiences of other biotechnology companies. In fact, Cetus's history reads like the history of the whole industry in microcosm.

Like the companies that followed its lead later in the decade, Cetus

Cetus

Dr. Peter J. Farley (left), Dr. Ronald E. Cape (right), and Dr. Donald A. Glaser (not shown) were the first to form a biotechnology firm in the early 1970s. Through their Cetus Corporation, they would ". . . take biology wherever it will lead us."

still generates most of its income from research projects done on a contractual basis for large corporations. In Cetus's case, three of those corporations—Standard Oil of California, Standard Oil Company (Indiana), and the National Distillers and Chemical Corporation—became major shareholders, collectively owning 49 percent equity.

According to industry analysts, therein lies the crux of Cetus —and its counterparts'—major problem. To this day, most biotechnology companies function like well-subsidized research laboratories, not like commercial enterprises that live or die on the performance of their products in the marketplace. They're service-oriented rather than marketing-oriented. Indeed, Cetus—along with most of its competitors—has few if any products to sell. Those products they do develop from their smorgasbord of scientific projects can't be marketed until the FDA confers its approval. Even then, most genetic-engineering firms' products probably will be marketed by the large *Fortune* 500-sized company that funded the research in the first place. Cetus and its counterparts' payoff will be a licensing fee plus royalties.

Therein lies a second problem: the lack of a narrow corporate focus. Cetus's research projects, for example, range from medical and cancer diagnostics and therapeutics to agricultural chemicals and the genetic engineering of plants.

All of this is not to say that gene-splicing firms stand no chance of making their founders rich. On the contrary, many of the founders, including the Cetus trio, have cashed in their equity through the centuries-old device of taking the company public. The process has already elevated many a struggling scientist into the citadel of the superrich. But for biotechnology firms, going public has its drawbacks, too.

A recent trend in the industry finds the scientist-founders being forced back into the labs where they function best. Replacing them in the executive suite are professional managers selected by investor constituencies. The new management's mandate is to take a more measured, short-term view of things and turn a loss-leader into a money-maker as soon as possible.

This trend reflects the truism that you can take a scientist out of the laboratory but you'll seldom get the laboratory out of the scientist, even if you award him an M.B.A. degree and shower him with cash. The average scientist-businessman hybrid remains forever a man of theory who has his hopes pinned on the future in the form of his next lab experiment. He has little regard for practical matters, such as whether anybody really wants or needs what he's concocting.

Under the scrutiny of investors, Cetus became embroiled in the type of managerial upheaval characteristic of entrepreneurial bioengineering firms with their large complement of scientific talent and dearth of strategic business thinkers. The management reshuffling at Cetus culminated in Dr. Farley's resignation in 1983. Not to worry. With his multimillion-dollar grubstake in Cetus stock, Farley—the man who had once bragged that he was "building another IBM"—announced plans to start a software company specializing in the health-care field.

Fellow founder-multimillionaires Cape and Glaser are still affiliated with the company. Ronald Cape was kicked upstairs to sit at the head of the table at Cetus board meetings. Donald Glaser, on the other hand, is the company's chief scientific consultant, the kingpin of an advisory board that reads like a *Who's Who in American Science.* The eminent Dr. Stanley Cohen, for one, is on Cetus's board.

During Cetus's early years, another company that's since become the flagship firm within the industry was also in its formative stage—inside a certain Robert A. Swanson's head.

In 1975, Bob Swanson was a twenty-eight-year-old would-be entrepreneur who was casting about for the right combination of ideas and expertise to spawn a bioengineering company of his own. He had a B.S. in chemistry and M.B.A. from MIT and he'd just resigned his partnership from the premier San Francisco venture-capital firm of Kleiner & Perkins.

In making the rounds of biochemists to solicit their advice, Swanson met, among others, the same Dr. Boyer whose discovery in 1973 had made the commercialization of genetics feasible in the first place. Dr. Boyer's attitude came as a pleasant surprise to Swanson. Compared to the average crusty academic, whose unfamiliarity with business approaches the xenophobic, Dr. Boyer seemed to hold a refreshingly opportunistic view of DNA's possibilities.

Their initial discussion occurred one Friday afternoon in January 1976. By May, Swanson and Boyer had $100,000 from Kleiner &

Steve Northrup/Black Star

Robert Swanson (left) teamed up with Dr. Herbert W. Boyer in 1976 to found Genentech, now the leading firm in the biotechnology industry. Here they stand in front of a "gene machine" or computerized chemical robot. The machine directs the chemical ingredients of DNA in a programmed sequence into a glass column where the new strand of DNA "grows."

Perkins to fund their partnership Genentech's maiden research project. It involved the synthesis of somatostatin, a hormone in the brain that regulates the secretion of other hormones. They'd chosen it because it has a health-care application that is low-volume but high-margin. They reasoned that a success in synthesizing somatostatin using gene-splicing techniques would surely impress potential investors and have a positive impact on public opinion.

It did. The scientific and business communities alike hailed their success with somatostatin as "a scientific achievement of the first order" that would one day render obsolete the conventional methods for obtaining human chemical substances. Those conventional methods include extracting the substances from human cadavers, a process that is painstaking and yields inadequate supplies.

Other Genentech syntheses of human and animal health-care products followed at a rate that eventually accelerated to several a year. There was human insulin in 1978; human growth hormone and thymosin alpha-one in 1979; human proinsulin (precursor to insulin), several subtypes of human leukocyte interferon, several hybrid leukocyte inferons, human fibroblast interferon, and bovine growth hormone in 1980; and human immune interferon, porcine growth hormone, vaccine for foot-and-mouth disease, human calcitonin, and human albumin in 1981.

The string of Genentech breakthroughs took the spotlight off biotechnology's dreaded liabilities and highlighted, instead, its latent benefits. Those benefits range from cures for cancer and the development of hybrid plants that resist insects without the aid of pesticides to the synthesis of enzymes that turn solid waste into useful products.

In the late 1970s, with Cetus and Genentech and the two hundred other embryonic gene-splicing companies scattered around the globe issuing press releases every second day describing their latest laboratory coup, the age of biohype was inaugurated. The era promises to last at least another decade.

It is an amazing growth period, unprecedented in the annals of science if not finance. It is a time when everyone from scientists to venture capitalists to private investors appears willing to sell his or her soul for a piece of the action. Forget the fact that biotechnology firms as a class have proven themselves far better at making miracles than money. During the peak of the frenzy in the early 1980s, no one seemed to care.

For a while, it looked like the fast-talking promoters of upstart bioengineering firms in league with skillful stock-market operators would make off with the largest chunk of the newfound biotreasure. It was certainly true that the shrewdest of the lot profited handsomely from everyone else's cupidity.

The corporate promoters went after the high-stakes gamblers, those wealthy individuals who, with the proper nurturing, could be induced to plunk down thousands during early rounds of venture-capital financing. As the more cynical hucksters put it, their pigeons were "investing in a professor" because the only way to value one of these unprofitable young companies was to count the number of Ph.D.s they employed.

If ever there existed a *caveat emptor* situation, this was it. But those rich investors who succumbed to the promoters' blandishments probably got what they deserved, since the sincerity of many biotechnology hucksters ranks right up there with that of the real-estate flacks who regularly beat the bushes for suckers dumb enough to invest in Nevada desert subdivisions.

Emblematic of Wall Street's biohysteria was the stampede that accompanied Genentech's initial public offering in 1980. Genentech's shares, issued at $35 apiece on the morning of October 14, 1980, peaked at $89 after twenty minutes of trading. At the close of the day, when the price settled at $71.25, Robert Swanson and Herbert Boyer each owned equity worth $70.2 million. All this activity, mind you, swirled around the stock of a company with an uncertain future that showed a minuscule profit of $51,802 on its balance sheet. Nonetheless, Genentech's securities experienced the most striking price explosion of any new issue in most of the participating stockbrokers' memory. It is a day now permanently enshrined in the record book of the American stock market under the heading "Crowd Psychology and the Gullibility of the Herd."

Inevitably the biobubble burst. In 1982, the industry found itself in the throes of a shake-out that caused a host of gene-splicing firms to close their doors forever and forced others to reorganize under the federal bankruptcy statutes. Among the latter group was Southern Biotech. The distressed company had set up, among other dubious practices, a subsidiary to handle its finances on Grand Cayman

Island, an infamous tax haven where transactions are not subject to U.S. banking regulations. As a consequence, Southern Biotech found itself being investigated by the FBI, the Justice Department's Organized Crime Strike Force, as well as a Tampa grand jury.

Of the scores of little biotechnology firms that went public in the early 1980s, Genentech is one that may yet live up to the excitement it generated in the public markets. In late 1982, Genentech became the first biotechnology company in the United States to announce it had received FDA approval to market one of its products without a prescription. Genentech's is not the first bioengineered product to hit the market anywhere in the world, however. A company in Europe got there first with a recombinant-DNA vaccine to treat diarrhea in piglets.

The Genentech product is a new form of insulin, called Humulin, that is synthesized from the human genes that code for insulin. The older form of the substance, which will remain the less expensive of the two for several years to come, is simply insulin extracted from the pancreatic cells of animals, usually pigs and cattle.

If the FDA approval for Humulin, which came in five short months, serves as a model for other bioengineered drugs, we will soon see a steady stream of them swooshing through the regulatory pipeline onto the shelves of the nation's drugstores. Since bioengineered drugs usually are replicas of substances manufactured in the human body, the approval process can move faster because the products represent fewer unknowns. In contrast, drugs synthesized from substances foreign to the body have a tendency to induce unpleasant side effects and allergic reactions.

A licensing and royalty agreement gives Eli Lilly, the Indianapolis pharmaceutical giant, the rights to manufacture and market Humulin. Genentech stands to make at most 10 percent of any profits from Humulin. But since that income will go straight to Genentech's bottom line, it will add to the war chest the company is amassing to underwrite its own bid to become another Eli Lilly—in short, a full-service drug company that handles everything from research through packaging and sales. To that end, Genentech maintains that once it receives the FDA go-ahead for its growth hormone and gamma interferon, it will muster internal resources to handle the marketing.

* * *

The hullabaloo surrounding Humulin was just the kind of publicity that in 1974 had focused the attention of Stanford University's patent officer on the profits to be made in the new industry. Quite by accident, the Stanford official happened to read an article in *The New York Times* about the wonders of the new science and its originators. One of those originators was, to the official's surprise, a Stanford employee.

Immediately, the patent officer swung into action and checked whether Stanford researcher Stanley Cohen had filed for a patent on his breakthrough methodology. He hadn't. In true absentminded-professor fashion, both Cohen and his colleague Herbert Boyer at the University of California had neglected to mention their discovery's commercial significance to the administrators at their respective schools.

If that quick-witted patent official hadn't moved swiftly, the recombinant DNA technique would have remained forever an open sesame to any scientist wishing to exploit it. As it was, the technology came within a week of being unpatentable because the one-year grace period between an "invention" and the filing of a patent application had nearly expired.

On behalf of Stanford as well as Cohen, Boyer, and the University of California at San Francisco, the officer filed for a patent. The original application was subsequently revised and refiled as two separate claims. Stanford requested patent protection for the gene-splicing *process* itself and also for the *products* resulting from the use of the process.

In December 1980—six months after the landmark United States Supreme Court ruling that new forms of life created in the laboratory could be patented—the patent on the process was granted. It gave Stanford, as the group's representative, the right to collect licensing fees from any bioengineering lab wishing to utilize the technique. By the spring of 1982, Stanford was receiving $10,000 a year from each of seventy-three licensees, a fee applied against royalties at the point any of the firms begin marketing products.

But the need to pay Stanford's tribute was called into question when the product patent was suddenly denied. It's a decision that Stanford is still appealing because, in many respects, it's the more important patent of the two.

In the midst of Stanford's battle to achieve this key product patent, several other technicalities presented themselves. It seems that a magazine article alluding to Boyer and Cohen's process was published slightly over a year before the Stanford University patent application was filed. With the article as evidence, critics maintained Stanford had not met the deadline for filing an application.

In addition, the U.S. Patent Office also questioned the authorship of the process itself. Robert B. Helling, who as a graduate student worked with Dr. Boyer and had his name included in the by-line on the official scientific paper announcing the technique, did not have his name included by Stanford on the patent applications. Helling is now a professor of biological sciences at the University of Michigan, and that school's lawyers are pressing his claim.

Inevitably, while the Stanford group waits and hopes, biotechnology companies are rushing into the void with their own patent applications. By the end of 1982, several thousand patents were pending on various processes and microorganisms, and hundreds had been issued. Genentech alone holds more than six hundred.

In a mad scramble to play it safe, the players in the biotech game are turning their industry into a patent attorney's idea of Nirvana. Why? Because it's a revolutionary technology and no one knows where it will lead.

"The methods developed by the biotechnology industry are going to be of crucial importance to many other industries," explains Dr. Zsolt Harsanyi, a former Cornell University geneticist, currently an E. F. Hutton vice-president and author of *Genetic Prophecy.* "What's happening now is that everyone is jockeying for position within the industry. As soon as it gets to the point where the products on the market make it worthwhile to sue, all the big patents are going to be challenged."

Litigating should be worth the gamble, since three out of four patents challenged in court are ruled invalid.

Japan may be ahead of the United States in robotics and tied for first place in the semiconductor field, but there's no contest in the realm of biotechnology. Even the second-tier American gene-splicers—such firms as Bethesda Research Laboratories, Collaborative Research, Inc., and Enzo Biochem, Inc.—are far ahead of Japan's entries, although the Japanese government is offering subsidies to help its native industry catch up.

Four firms, either resident in the United States or with American origins, have soared ahead of the competition to capture worldwide leadership positions in this teeming industry. In addition to Genentech and Cetus, they are: Genex of Rockville, Maryland, and Biogen N.V., headquartered in Switzerland and with a laboratory in Cambridge, Massachusetts.

Genex Corporation is the progeny of yet another marriage between a scientist and a businessman. It is the construct of Dr. J. Leslie Glick, former chairman of the physiology department at the State University of New York at Buffalo, and Robert F. Johnston, an M.B.A. with fifteen years' experience as a troubleshooter for entrepreneurial companies seeking financing.

Like its counterparts, Genex subsists on contract research for major corporations in the industries most likely to realize substantial benefits from the new technology—chemistry, food processing, agriculture, mining, and waste treatment. Also typical is the fact that Genex's eighty-odd corporate clients and/or joint-venture partners span the globe. Besides American corporations (e.g., Bristol-Myers, Bendix, Koppers, National Cancer Institute), Genex's business relationships encompass as well a fair number of Japanese (e.g., Mitsui Toatsu Chemicals, Yamanouchi Pharmaceutical Co., Green Cross Corporation) and European companies (Kabi-Vitrum, Fortia, Schering AG).

Genex, once again, exemplifies how corporate America has infiltrated this emerging industry. Genex is backed by two large equity partners: Koppers Company, whose ownership peaked at 45 percent, and InnoVen, a group of holding companies owned by Monsanto and Emerson Electric Company. InnoVen's percentage peaked at 22 percent. But Glick and Johnston also retained their fair share of equity —enough to make each of them worth $16 million by 1983.

The scientist-academic reborn as a *nouveau-riche* Croesus is becoming a common phenomenon in the biotechnology field. Dr. Walter Gilbert, who left a tenured chair at Harvard to head Biogen, typifies the compromises forced on professors when riches beckon.

Kevin Landry, a general partner at the Boston venture-capital firm of TA Associates, was actually the catalyst for the formation of

Biogen N.V. in 1978. When Landry's search for an investment opportunity in genetic engineering turned up no likely prospects, he decided TA Associates would have to engineer its own. "We can't wait for these things to fall into our lap," Landry explained. "You have to make your own opportunities."

Genex

Robert F. Johnston (standing), an experienced troubleshooter for entrepreneurial companies seeking financing, and Dr. J. Leslie Glick of the State University of New York at Buffalo are the prime movers behind Genex Corporation. Despite its problems, Genex still is one of the country's foremost genetic engineering firms.

Landry, with his excellent contacts in Cambridge scientific circles, had little trouble drumming up interest for his proposed firm. In fact, a number of respected biochemists agreed to serve as consultants, including Dr. Gilbert. Gilbert's cooperation was key, for, among other credentials, he shares a Nobel Prize in chemistry for a method he developed to determine the chemical sequence of DNA.

With Gilbert and his associates eager to participate, Landry established a major research lab for Biogen in Cambridge. But he claims he headquartered the company in Geneva, Switzerland, to exploit the relatively untapped reservoir of European talent in the molecular biology field.

Dr. Gilbert's affiliation with Biogen may have been a boon for Kevin Landry and TA Associates, but back at Harvard, the news got a lukewarm reception. Indeed, all it did was force Harvard administrators to confront a growing academic problem, one begging for resolution not only at the country's oldest institution of higher learning but also at the approximately one hundred other major U.S. research universities.

The problem highlights the increasing reliance of American corporations on universities for applied research work. The issue—one of conflict of interest—dates back to the beginning of this century, when the country's science-based industries and technical-school educators first started to cooperate to promote the cause of research in general. The deal the industrialists and educators struck then is still operative today. In short, to promote America's technological progress, companies would supply the research funds and universities would supply the brainpower.

Over the years, the conflict-of-interest problem flared up periodically and is doing so again in the 1980s. Most of the molecular biologists needed in the genetic-engineering field are resident in universities. It's a small pool of talent that many university administrators do not want to see corrupted by money.

A number of observers, including some outspoken scientists, are simply bemused by university administrators' holier-than-thou attitude. "The values which we see out there in society permeate science," says Richard Goldstein of the Department of Microbiology and Molecular Genetics at Harvard Medical School. "And I don't think that the scientists are making any other very different types of decisions from what most people make in 1981; in terms of their life-style, the types of positions they take, and how they keep their jobs, and where they take funds, and what compromises they're going to make. People make many more compromises in 1981 than they did in 1968."

Biogen, Inc.

Dr. Walter Gilbert, the Nobel Prize-winning biochemist shown standing above, left a tenured chair at Harvard University to head Biogen. This successful gene-splicing firm was the idea of venture capitalist Kevin Landry, who assembled the financing and recruited the scientists, including Dr. Gilbert.

In the best of all possible academic worlds, of course, a professor's goal is supposed to be the pursuit of basic research, wherever it may lead. That pursuit, by definition, implies that a scientist will share his findings with his peers in the spirit of free and open exchange. Therein lies the conflict: How can a university scientist be expected to keep his motives pure when the value of his company stock—awarded him in lieu of a consulting fee—depends on whether the end result of his lab work yields a commercially viable product? It also depends on whether he keeps his mouth shut about his discoveries when queried by his academic colleagues.

On the other hand, scholars are entitled to make a decent living, too, and, in most cases, that's not easy on an academic salary. Supplementing inadequate salaries is precisely the reason why most university scientists began renting their brains to commercial enterprises in the first place.

Although some forty or fifty Harvard professors were partners in or consultants to various biotechnology firms by the early 1980s, the Gilbert appointment, because of his stature and the success of Biogen, brought the issue into the sharpest focus. In November 1981, Harvard's faculty adopted new guidelines regulating the outside activities of full-time Harvard professors. In essence, the rules make it clear that dual allegiances will not be tolerated. Twenty percent is the maximum amount of time a faculty member can now commit to profitmaking activities other than teaching or research at the university.

Harvard's stern approach catapulted Walter Gilbert out of his chair at the university into a lucrative job as chief executive officer of Biogen. The financial payoff came in 1983 when the company went public, giving Gilbert's equity a market value of $8.4 million. He's the company's largest individual shareholder.

While Faustian overtones permeated the academic-freedom debate back East, no such puritanical images appeared to blacken the sunny optimism of the principal West Coast schools involved with gene-splicing. In fact, while Harvard was purging its institutional soul of the evils inherent in profit, officials at Stanford University and the University of California at Berkeley were working out a deal that would help their respective institutions profit from the new technology.

In a novel arrangement, the two schools set up a joint nonprofit Center for Biotechnology Research, whose start-up capital was provided by six huge corporations. In turn, the center holds 30 percent equity in a commercial venture called Engenics, Inc., which will market any useful products emanating from the center's lab. Moreover, the same six corporations that underwrote the establishment of the center also own an aggregate 35 percent equity in Engenics. The universities' officials say any capital gains or dividends Engenics generates, the center will channel back into research at designated departments within the two schools.

Stanford and UCB administrators maintain that this unusual setup avoids the potential conflicts of interest that might arise if the schools themselves were shareholders in companies such as Genentech, created by faculty members.

Whether or not the originators (Cohen, Boyer, and their respective schools) of the recombinant-DNA technique get their patent, at least they filed for one. The two European immunologists who in 1975 devised the process creating hybridoma technology, a sister technology to gene-splicing, weren't so farsighted. They did nothing to cash in on their "invention," leaving their technology wide open to all comers.

But monetary dreams have since seized hold of the scientists at the Wistar Institute of Anatomy and Biology in Philadelphia. They refined the basic hybridoma technique and received two patents that many expect to be challenged in court. In the meantime, a number of entrepreneurial companies are also hell-bent on eking out whatever commercial gain they can from this startling process with its panoramic potential.

When they made their pioneering discovery, Dr. Cesar Milstein and Georges Kohler, working for Britain's Medical Research Council in Cambridge, were searching for a better way to collect and purify antibodies. Antibodies are the disease-fighting weapons manufactured by the white cells in the blood of all higher-order animal species. The old way of harvesting them is arduous, inefficient, and imprecise. It's done by injecting a foreign substance, or antigen, into a test animal, waiting for the creature's natural immune system to produce antibodies, and then separating out the target antibodies from the creature's blood. Unfortunately, this method also reaps a menagerie of other, unwanted antibodies. Isolating the desired ones is very difficult.

Milstein and Kohler's innovative process is not only fecund, but also the end product is one antibody instead of a whole zooful. Their method calls for the creation of a hybrid cell, termed a hybridoma, by means of "cell fusion"—in short, merging normally short-lived natural antibodies with long-lived tumor cells. Due to the rapid cell division characteristic of cancer cells, an army of ultrapure antibodies of a single type—hence the term "*mono*clonal antibodies"—is quickly produced.

What are they good for? These monoclonal antibodies are proving to be exceptionally useful when incorporated in medical diagnostic kits, where they work to identify specific diseases and other abnormal bodily conditions such as pregnancy, allergy, and anemia. Someday, scientists also expect to deploy monoclonal antibodies as a kind of magic bullet to quell the same diseases they help detect.

The entrepreneurs who are harnessing the power of the body's natural immunochemistry in this manner in the hope of garnering huge profits include a large quotient of venture capitalists.

Hybritech, bucking to capture the leadership position within this subindustry, is a prime example of a firm assembled by an aggressive venture capitalist-turned-entrepreneur. He's Brook H. Byers, a partner in the same San Francisco venture-capital firm that reaped the outsized return on Genentech.

In the age of high tech, a good venture capitalist has to be an instant expert. Should business opportunities open up in some arcane technical discipline—monoclonal antibodies, let's say—our venture capitalist must be able to absorb (overnight, if need be) enough knowledge to play devil's advocate to the real experts. The real experts, of course, are the entrepreneurs who, in the ideal capitalist world, plead for seed money to underwrite their innovative notions. Then it becomes the venture capitalist's job to separate the harebrained from the truly inspired.

No one knows all this better than Byers, who says he gets some of his best investment ideas by indulging his interests. Medical discoveries are one of them. When Byers first read about fusing cells to produce pure antibodies, he was intrigued. The next day, he bought a book of essays on immunology and boned up further. What he learned excited him so much he hired a researcher to sift through the scientific literature on the subject. Within a month, Byers was convinced that hybridoma technology represented a breakthrough of the greatest significance, one he wanted his firm to exploit despite the fact that no wild-eyed microbiology professors, pushing monoclonal antibodies, had come begging for the requisite handout.

Byers signed on two University of California at San Diego research scientists—Howard C. Birndorf and Ivor Royston, M.D.—as his "cofounders." Next, he made a new-business presentation to his own firm's partners. He came away with a $1.7 million nod of approval for his fantastic notion. Hybritech, Inc., of La Jolla, California, opened its lab for business in 1978, with Byers acting as the company's first chief executive officer. It's a post he has since relinquished to a professional manager.

Hybritech quickly moved out ahead of its competition, firms bearing such names as Monoclonal Antibodies, Inc. (Palo Alto). Hybritech was the first monoclonal-antibody producer to win FDA approval for a diagnostic test kit, and by 1982 it actually had four different TANDEM Assay System kits on the market. The company plans to develop some fifteen more individual kits, which are used by physicians to help diagnose and monitor various diseases. In addition, Hybritech is accelerating its therapeutic-product research to ensure that it's among the first firms allowed to undertake clinical trials for a monoclonal-antibody cancer treatment.

However, when it comes to drugs engineered by means of hybridoma technology, Hybritech does have several serious entrepreneurial competitors. Cytogen Corporation and Centocor Inc., which are peopled by some of the same Wistar Institute scientists who have their names on that institute's two hybridoma patents, are also considered likely contenders in the race to use monoclonal antibodies to cure cancer and other deadly diseases. Then there's Genetic Systems Corporation of Seattle, a firm that in 1984 received a patent on a process that raised the production levels of human monoclonal antibodies to commercial levels for the first time. Using this process, scientists can produce twenty to one hundred times greater concentrations of antibodies than they could previously.

When Hybritech went public in 1981, Brook Byers only had direct ownership of stock valued at $200,000. In contrast, Dr. Royston's securities were valued at $4.7 million. But don't weep for Brook

Byers. Through his partnership stake in Kleiner, Perkins, Caufield & Byers, he also shares in equity worth a handsome $28 million.

David Padwa is an unlikely person to head a Denver seed company that doubles as one of the nation's handful of horticultural genetic-engineering firms. An ivory-tower biochemist Padwa is not. Quite the contrary. He's a practical man of the world, a lawyer by training and a computer expert by dint of his first entrepreneurial venture back in the 1960s.

Padwa sold that original company, Basic Systems, to Xerox in 1965 for stock worth $6 million. That transaction made him rich enough to take a few years off to indulge some fantasies and see the world. But by the early 1970s, Padwa was bored and determined to make a second fortune.

In his adobe abode in Santa Fe, New Mexico, Padwa, the perpetual student, began his search to identify industries with high growth potential. After burying himself in books and periodicals for several months, Padwa, the visionary, surfaced with the conviction that a world food shortage loomed just over the horizon. That was the problem. Padwa's solution was to deploy the revolutionary new gene-splicing technology to manipulate and improve the genetic makeup of plants.

Hybridization or "cross-breeding" is the traditional way botanical scientists create better plant strains. Utilizing this evolutionary process, scientists breed generation after generation of a plant to isolate a mutant with the desired trait. Granted, this method has resulted in a number of useful food products, but it can take years before it quite literally bears fruit. The process is largely a matter of trial and error, superhuman persistence, and blind luck.

In contrast, the more rifle-shot techniques of the biotechnologist are far more efficient. Their advantages are twofold: Genetic-engineering techniques speed up nature's rather leisurely timetable. Not only that, but also DNA theory coupled with the gene-splicing techniques give plant breeders, for the first time, a thorough understanding of why a certain trait expresses itself.

If the genetic manipulation of plant DNA lives up to its promise, a decade from now we should bear witness to such wonders as self-fertilizing crops that will be much cheaper to grow because they require little or no energy-intensive nitrogen fertilizers . . . crops designed for higher yields and greater nutrition content . . . warm-weather plants modified to thrive in harsh climates . . . fruit with a better water-to-pulp ratio . . . and plants that are more resistant to blight.

Padwa refined his plans: He would found a company that would peddle the farmer's staple—seeds—and use the profits to underwrite biotechnology-based crop research. It made perfect sense to Padwa. "After all," he says, "anyone in the seed business is already in the genetic container business."

A string of his wealthy acquaintances thought it made perfect sense, too, and invested millions. Among those who chose to gamble with Padwa's Agrigenetics Corporation are Henry Ford II and Stewart Mott, the car scions; Max Palevsky, the Los Angeles computer magnate; William Weyerhaeuser of timber origins; and such high-visibility Europeans as the British and French Rothschilds, the Flick family of Germany, and the Bonnier family of Sweden.

In 1975, Padwa began buying up the eleven or so seed companies that would form Agrigenetics' nucleus. Everything went according to the prospectus until such industrial behemoths as Monsanto and Atlantic Richfield started channeling millions into plant R&D.

They represented one threat. The other appeared in the guise of a tiny San Carlos, California, firm named International Plant Research Institute (IPRI), the brainchild of three high-powered scientists named Martin A. Apple, Ph.D.; his uncle, William S. Apple, M.D.; and Robert E. Abrams, M.D. IPRI as well as Genentech, Cetus, and several other San Francisco Bay Area gene-splicers were causing pundits to term the region Sili*clone* Valley—and not just because of their mammalian DNA research. Many of that valley's biotechnology companies were also diversifying into various aspects of botanic R&D.

To match and to try to exceed the competition's horticultural R&D budgets, Padwa organized a series of limited R&D partnerships that netted Agrigenetics more than $50 million. This strategy transformed Agrigenetics into one of the best-financed biotechnology firms doing any kind of research. And Padwa is well on his way to assembling his second fortune, of a magnitude far greater than his first. In 1984, Padwa sold out to Lubrizol Corp. for an undisclosed amount.

PART TWO

The Computer Renaissance

2 Hardware Horizons: *The Computer Manufacturers*

Just as Henry Ford was not the first manufacturer of automobiles, neither were the founders of Apple Computer the first producers of microcomputers. Among Apple's predecessors were Micro Instrumentation and Telemetry Systems (MITS), IMSAI, and Vector Graphic.

Through mail-order ads in amateur science magazines, these early firms sold computer kits to hobbyists who assembled the state-of-the-art electronic components into a finished product for use in their homes. But even though such kits were surprisingly popular and sent to customers COD so these companies' seed money requirements were minimal, only Vector Graphic survived as an independent entity.

Vector lives on because its three founders—two bored housewives with no background in engineering or computers, and one of their husbands—had the prescience to move with the onrushing electronic tidal wave into the marketing of fully assembled microcomputers. In contrast, MITS and its line of Altair computers was acquired by the Pertec Computer Corporation in 1977, making its principals, Ed Roberts and Eddie Curry, some $6 million richer in Pertec stock. And IMSAI perished in bankruptcy court in March 1979. But IMSAI was no loss to its founder, William H. Millard, whose second business was already well on its way to becoming the country's largest chain of computer stores.

Meanwhile, at the peak of the computer-kit fad of '75 and '76, two other ambitious young men were hard at work in a California garage designing one of the world's first plug-'em-in-and-they-work microcomputers. At the time, Steven P. Jobs was just twenty-one and his buddy Stephen G. Wozniak—nicknamed "the Woz"—a mere twenty-five; and Apple Computer, Inc., was nothing more than a psychedelic figment of their fertile postadolescent imaginations.

The pair had met in the late 1960s, when Jobs was in junior high and Wozniak was in high school in Los Altos, an affluent community in the heart of Silicon Valley. As a consequence of their natural technical bent and the counterculture milieu where they lived, Jobs speaks for both when he claims, "I grew up with Bob Noyce [cofounder of Intel] as a cultural hero, along with Camus, Sartre, and Bob Dylan." Indeed, the duo's anti-establishment tendencies may be what brought them together in the first place.

Jobs and Wozniak had each put in a year or two in college before they plunged into their life's work, noodling with sophisticated electronic gear for fun and profit. Wozniak was a designer at Hewlett-Packard and Jobs on the payroll of the video-games genius Nolan Bushnell at Atari when they first teamed up after hours. Their initial product was a "blue box," the bane of the telephone company's existence because the compact boxes can be hooked up to any phone anywhere, allowing users to to make long-distance calls free all over the world. When AT&T began to pursue malefactors, the pair decided to channel their ingenuity into more legitimate endeavors.

Utilizing that newly created electronic wonder—the microprocessor—as the key component, the two spent six months in one of the Jobs family bedrooms and the garage designing the prototype for a typewriter-size microcomputer. Through word of mouth they soon had orders for fifty of these elementary microcomputers that would later be dubbed Apple I. But where to raise the start-up capital to buy the materials to build them? By selling Jobs's Volkwagen bus and Wozniak's Hewlett-Packard scientific calculator, they scraped together the necessary $1,300.

Their makeshift production line consisted of the two of them and Jobs's sister Patty. The end product was little more than an assembled, printed circuit board with no box around it. Such accessories as a keyboard and video monitor had to be purchased elsewhere. Even so, their microcomputer was so popular at their asking price of $666.66 that gross sales soon amounted to more than $100,000.

With money like that rolling in by 1976, Jobs, a workaholic promoter, started to think about creating an ongoing business of magnitude. The Woz, who at age thirteen had built his first computer and captured first prize in a science fair, was more interested in creating an aesthetically perfect microcomputing machine regardless of the monetary reward. Fortunately for them, their goals were not mutually exclusive.

Jobs made ample use of the Silicon Valley grapevine in spreading the word about their brainstorm—and in soliciting technical advice that he relayed to the Woz about how to improve the product. One person Jobs approached for financial backing was venture capitalist Don Valentine, who dismissed him as "a renegade from the human race."

It's true, Jobs did look more like a refugee from Haight-Ashbury with his dark, shoulder-length hair and sandaled feet than a hotshot inventor soliciting serious venture-capital funding. And the Woz—well, according to one friend, "He looked like a Steiff teddy bear on a maintenance dose of marshmallows. Still does."

But Valentine did mention the meeting with Jobs to a young investor named A. C. Markkula (pronounced MAR-koo-la), Jr. "Mike" Markkula had worked for Fairchild Semiconductor and Intel, earned a reputation as a marketing wizard, and then retired at age thirty-two. In a short time, he'd made $1 million through astute investments.

Markkula recalls that he only intended to help the two struggling boy-entrepreneurs write a business plan. Instead, he ended up giving them $250,000 of his own money and taking over as chairman of the nascent venture. It incorporated as Apple Computer in January 1977.

The whimsical name "Apple" came about by accident. In frustration, Jobs seized on it because he happened to be eating one during the long-winded debate over the company's name. Only in retrospect did the name and the company's bright, multicolored apple logo strike these two overgrown flower children as the ideal, simple, and decidedly jargon-free symbol of everything they wanted the company to stand for.

In May 1977, Apple startled the computer industry and amazed the world at large when it introduced the Apple II, a fully assembled, programmable, "friendly" microcomputer. In the company's impressive advertisements, the Apple II was dubbed a "personal computer," emphasizing the idea that it was intended for one person's use. The advertisements also pointed out that the lay personal-computer user no longer had to know how to solder, wire, assemble, and debug his own machine. Now he or she could buy one right off the shelf of a store and start computing. Moreover, the Apple II was the first microcomputer whose data storage device was a floppy disk system rather than a slow and inefficient cassette recorder.

But the most amazing thing about the Apple II was something else —its brainpower. It was equivalent to an IBM Model 1401, a cum-

Apple Computer

Steven P. Jobs is pictured with the Apple II. This best-selling microcomputer pushed Jobs's personal net worth over the $250 million mark when the company he cofounded, Apple Computer, Inc., went public in December 1980.

bersome mainframe that was IBM's top-of-the-line model until 1966, predating Jobs's and Wozniak's handiwork by ten years.

Indeed, the 1960s and 1970s have seen the most prodigious price-performance improvements in computer technology of all time. IBM's mainframe Model 1401 had thirty-two thousand words' worth of main memory. With its peripherals, it cost about $120,000 (about $340,000 in 1981 dollars). It filled a six-hundred-square-foot room that had to be air-conditioned and required raised flooring. By 1978, a $3,000 Apple computer with sixty-four thousand words of memory did the job of two 1401s—for 3 percent of the price and 1 percent of the space, space that needed no air-conditioning or other special treatment.

Apple Computer

Stephen G. Wozniak, a high-school buddy of Steve Jobs's, designed the Apple II. "The Woz" owned shares worth more than $125 million when Apple Computer went public.

* * *

At about the time the Apple II and other fully assembled microcomputers were appearing, a number of entrepreneurs were exploring the notion of computer retailing as a possible avenue to wealth.

The first entrepreneur had rushed into the void as early as July 15, 1975. That was the date when computer buff Dick Heiser opened America's—maybe the world's—first computer specialty store, in Santa Monica, California. He gave it a straightforward name—"The Computer Store"—and stocked it with everything a microcomputer freak would ever need to feed his compulsion. In those days that meant $500 computer kits, components, and technical books.

When fully assembled microcomputers hit the market a year or so later, Heiser's imitators began appearing. And by the end of 1978, when the microcomputer fad had transformed itself into a minor revolution, about seven hundred computer stores were spread around the country. But the hotbed of them remained sunny California, where many American trends seem to blossom first.

In 1976, a year after The Computer Store proved there was a need, Bill Millard founded his second start-up venture, ComputerLand. With IMSAI microcomputers to sell and few outlets in which to do it, Millard immediately grasped the need for retail stores. After all, fielding a sales force was expensive—so expensive, in fact, that goods sold directly to consumers via sales representatives were generally marked up a minimum of 40 percent just to pay for the individualized attention. Clearly, an item had to carry a high price tag to make it worth the sales rep's time and effort.

But due to technological advances, the price of microcomputers was spiraling downward rather than upward. And the market for them appeared to be cost-conscious individuals, not well-heeled corporations. So offering microcomputers in retail outlets, where customers could kick the tires of several competing systems before making a choice, seemed logical.

ComputerLand opened its first store in the San Francisco Bay Area. A year later, Millard started selling franchises. By the end of 1983, ComputerLand had more than 580 franchised stores, and Millard's privately held company had an estimated market value approaching $1 billion.

Already Millard's 94 percent equity in ComputerLand ranks him among the wealthiest people in America. The next largest shareholder, with 5 percent, is Millard's partner Edward Faber—that is, unless two former IMSAI employees and an ex-ComputerLand consultant win their respective lawsuits filed against Millard in an attempt to share in his success.

By 1979, two important things made Apple, for a time, practically invincible against its growing list of competitors. First, Apple was embraced by a horde of computer freaks, more commonly known as "hackers," "nerds," and "droids." For the most part, they were middle-class adolescents and antisocial young adults obsessed with the new electronic wonders and possessed with a knowledge of programming.

The second thing involved the avant-garde in the business community. These alert, fast-track managers began to notice that, for a fairly trivial sum, they could buy a machine that would perform many of the functions heretofore relegated to the corporate computer department—and, what's more, it could do them better, faster, and right at their own desks.

Needless to say, a computer without a set of programs—instructions telling it what to do—is about as useful as a television set would have been in the year 1800. But to Apple's great advantage, all those computer freaks who'd bought all those Apple IIs in 1977 and 1978 wrote hundreds of programs, the stuff that's called "software" (as opposed to the computers themselves, which are called "hardware"). They wrote all kinds of programs, ranging from downright silly to extremely useful. They wrote game programs, electronic file cabinet programs known as "data bases," word processing programs, mailing-list programs—and, eventually, a simulated spread-sheet program that was destined to put Apple on the commercial map.

In 1978, a Canadian immigrant, a Harvard Business School graduate, and two MIT alumni joined forces in Cambridge, Massachusetts, to create and market a program called "VisiCalc" to be run on Apple IIs. VisiCalc is, in essence, a program that completely automates everything an M.B.A. is trained to do. It makes it possible for any businessperson to perform incredibly complex analyses with absolutely blinding speed. A rudimentary knowledge of high-school algebra is all that's required. VisiCalc, more than any other program, made the Apple the darling of the executive suite.

Until the advent of the Apple and VisiCalc, most businesspeople were simply frustrated by computers. Typically, company computers were headquartered in some off-limits place controlled by an army of bureaucratic empire-builders. To the average manager, these remote computers were little more than oversized numbers-crunchers spewing forth voluminous statistical reports that wrenched your back when you tried to lift them. Up to this point, very few computers—whether minis or mainframes—were made available to hard-working managers to analyze specific departmental problems. Rather, they were used for large-scale, corporatewide purposes such as payroll, accounting, order processing, and inventory control.

These frustrated managers represented what the business school professors would view as a classic, unmet market. Apple IIs—coupled with all that computer-freak software—met this need. Executives began buying Apples en masse, paying for them out of petty cash or their own wallets, just as they had with hand-held calculators a few years before.

With the cry "Processing to the people!" a guerrilla computer movement ignited. Suddenly all sorts of large companies with elaborate mainframe computer systems discovered their employees were ignoring the central electronic data processing department. It was an electronic do-it-yourself craze. Merrill Lynch found Apples on trading desks. Xerox learned its most successful sales managers refused to use the corporate computer system because they could forecast their sales faster and more accurately on an Apple. General Electric lost count of the number of Apples in its far-flung offices. So did others: CBS, Ford, Citicorp, Manufacturers Hanover, International Paper. The list grew longer every day.

Things changed fast once the Apple operation moved out of its garage. That first year, gross sales zoomed to about $7.8 million. Markkula, a slight, soft-spoken, conservative-looking man of Finnish descent, managed to raise $600,000 from a group of prestigious venture-capital sources that encompassed the Rockefeller family's Venrock Associates, Arthur Rock & Associates, and Teledyne founder Henry Singleton. In addition, he secured a credit line from the Bank of America. In fact, he found soliciting money for Apple so easy that he boasted he could raise another $8 million from the same sources if need be. But because of Apple II's explosive sales, no more was needed. Retained earnings financed the company's early growth just as profits had sustained Henry Ford earlier in this century.

There were other similarities between the Ford and Apple stories. Both had a good, dependable, no-frills product that they merely assembled from components made by others. Both priced their products low enough to have the broadest possible appeal. And both went for broke.

Markkula reasoned that a fragile new company such as Apple eventually would be crushed if it simply tried to carve out a nice little niche for itself in a mass consumer market as potentially large as the one for affordable microcomputers.

"Just aiming for a 10 percent market share is not viable," he said in a published interview in 1979. "We anticipate extreme rates of growth. We have to dominate the business or go bankrupt trying." In another interview, he said his long-range goals were to build a company that would still be around in the year 2000 and to crack the *Fortune* 500 list in the 1980s.

By 1981, the Apple team had fulfilled one of Markkula's goals. The feisty Apple had garnered a dominant market share competing against the industry's giants—Texas Instruments, Tandy's Radio Shack, Commodore International (a large percentage of its microcomputers are exported to Europe), and IBM, which had just entered the fray in August of that year—as well as more than a hundred pint-size entrepreneurial competitors with names such as Altos, Convergent Technologies, Cromemco, and North Star.

In 1982, Apple made good on another of Markkula's goals. With sales of $583 million, it slipped into slot No. 411 on the *Fortune* 500 list. Apple Computer is the first company ever to achieve *Fortune* 500 status after only five full years of operation.

But by 1983, with practically all the established mainframe and minicomputer companies producing micros and muscling their way into the marketplace with Darwinian ruthlessness, Apple Computer was forced to abdicate its overwhelming market lead. In that year Apple Computer slipped back to second place in market share behind IBM, a poor record compared to that of Ford Motors earlier in this century. But, then, times have changed.

In contrast, Henry Ford's hegemony lasted from the day the first four-cylinder, twenty-horsepower Model T rolled out of his plant on

October 1, 1908, until well into the 1920s, when the public finally tired of the company's one-model offering.

But for the first four years of its existence, Apple Computer's sales trajectory looked almost as spectacular as old Henry's: By the outbreak of World War I, Ford Motors had captured about half of the market for all new cars in the United States—and new-car sales were fast emerging as one of the country's major industries. In 1900 there were only 13,824 automobiles registered in the United States. By 1950, there were forty-four million.

In the same way, by the year 2000, all of us should be fully acclimated to the little micro's hums and whirs. This is an entirely different noise from the gentle clicking of electromechanical relays that characterized the 1944 Mark I computer's mental processes, reminding one bystander of the sound made by a roomful of ladies knitting. In our lifetime (1950–2000), if the most optimistic prognostications are correct, we'll witness computer growth equally astounding—from ground zero to some one hundred million in the year 2000 in the United States alone. Of that hundred million, 90 percent probably will be of the micro variety and used by individuals at home and work.

For all their congruences, the Apple and Ford stories diverge sharply when it comes to a discussion of how many people racked up personal fortunes from owning equity in the two companies. Henry Ford chose to keep the Ford Motor Company all to the Fords. He bought out all his minority stockholders in 1919, and thereafter the Ford family retained 100 percent ownership. The family retains control of the company to this day via an elaborate tax-avoidance scheme engineered by a Wall Street investment banker named Sidney Weinberg, former chairman of Goldman, Sachs.

Again, times have changed. Now the conventional practice is to take a company public as soon as it seems feasible, sometimes before the company has even marketed its first product. Apple Computer's public offering in December 1980 showered the company with $90 million in cash. It also made some forty Apple employees and a handful of the Woz's relatives and close friends millionaires, its founders centimillionaires, and its enthusiastic investors several million dollars wealthier.

At the market high in 1981, when the number of Apple IIs sold had reached 350,000 and annual sales topped $335 million, the newly minted Midases lined up as follows:

TABLE 6

Shareholder	*Shares Owned at First Public Offering (in millions)*	*Value at 1981 Market High of $34.50 (in millions)*
Steven P. Jobs (founder)	7.54	$ 260.13
A. C. Markkula, Jr. (founder)	7.03	$ 242.54
Stephen G. Wozniak (founder)	3.99	$ 137.66
Peter O. Crisp/Venrock (investor/director)	3.80	$ 131.10
Michael M. Scott (company officer)	2.81	$ 96.95
Dr. Henry E. Singleton (investor/director)	1.20	$ 41.40
Arthur Rock (investor/director)	.64	$ 22.08
Charles O. Finley (investor)	.35	$ 12.08
Philip S. Schlein (investor/director)	.11	$ 3.80

Newfound wealth has a way of magnifying people's basic character traits. That's certainly been the case with Apple's founding trio. Today, Jobs has cut his hair to an acceptable Wall Street length but he's still the same old workaholic at the company's helm. By nature he's a promoter and still pontificating on quasicounterculture themes even though his audience is now a group of button-down securities analysts.

The Woz got tired of being tagged a "college dropout," divorced his first wife, Alice (worth some $35 million to $40 million in Apple stock), married a second, and returned to the University of California at Berkeley under an assumed name to get his bachelor's degree.

The assumed name was so he wouldn't intimidate his computer science professors, although he conceded their courses were challenging. At this writing he is back at Apple Computer, where he's in charge of upgrading the Apple II, still the company's best-selling product.

Wozniak also unloaded some of his millions over the 1982 Labor Day in San Bernardino County, California, where he staged "The US Festival." It was billed as a celebration of song and science—actually rock and the microchip. Wozniak said he wants the 1980s to be thought of as the "us decade," erasing the me-me-me myopia of the 1970s.

Markkula is still a company officer. Off-hours, he's engaged in a running feud with his horsey neighbors in ultraposh Woodside, California. The issue: the regular deposits of horse droppings on the unpaved access road to his house.

The concept of a life cycle is as endemic to industries and companies as it is to plants and animals. Indeed, growth, competition, and consolidation characterize life on this planet.

An industry/company—or a human fetus, to use a more tangible example—begins life slowly because it's starting from a zero base. Even though it may be doubling in size, such small-percentage changes in size won't look like much for a while. But as the fetus continues to grow and gain momentum, it eventually becomes large enough that any percentage change appears significant. Then the fetus—or embryonic industry/company—is said to be in a period of rapid growth.

In the case of an industry or company, this expansionary phase finally ends when the entity begins to experience limits—the technology driving the industry or company reaches its ultimate potential, or a new competitor supplants it, for example.

After fifty years at full throttle, the automobile industry only recently embarked on the downward slide of its growth curve.

In 1899, the first year that the definitive *United States Census of Manufacturers* published auto industry statistics, there were some thirty American car companies. By 1908 the industry was entering its rapid-growth phase and there were 515 separate firms manufacturing cars. Competition was intense.

Many of the early auto entrepreneurs had come from related fields.

They were former wagon builders (Clement Studebaker's associates), bicyclemakers (Alexander Winton), plumbing suppliers (David Dunbar Buick), and axle manufacturers (Harry C. Stutz); and their vehicles bore such names as Stanley Steamer, Peerless, Pierce-Arrow, Owen Magnetic, and Stevens-Duryea. They had been drawn to the industry by the seemingly easy entry requirements—the relatively low initial capital outlay combined with the promise of making a personal fortune from a product that was proving it had mass appeal.

But the disappearance of most of those 515 companies when the industry entered its high-growth phase was part of a predictable capitalist pattern. It wasn't long before the mechanic-entrepreneurs mesmerized by nuts and bolts were overtaken by the fast-talking promoters obsessed with properties and profits.

The promoters went shopping with a vengeance for likely automobile companies to merge into combinations. The net effect was an industrywide consolidation. The most infamous consolidator was William C. Durant, who merged the Buick and Olds companies and several others under the charter of a New Jersey holding company, then lost and regained control of the holding company twice before finally abdicating to a more conservative management group controlled by du Pont and Morgan interests.

By 1915, Durant's General Motors and the Ford Motor Company had captured about 50 percent of the entire automobile market, with Ford producing and selling nearly five times as many cars as GM. By the 1960s, the "Big Two" and a horde of tiny competitors had transmogrified into the "Big Four" with no competitors. And they almost became the "Big Three" in the late 1970s when Chrysler threatened bankruptcy.

Compared to the automotive industry, the computer industry is still young and vigorous. But it's an industry divided into distinct segments, each in a different stage of its life cycle.

One commentator uses a jungle metaphor to describe the present state of the computer industry:

The industry is "rich, alluring, competitive—filled with wildlife vying for resources and territory," he writes. The dominant denizens, the mainframe computers, are as formidable as elephants and just as large. The smaller minicomputers resemble nimble, tree-swinging orangutans, adept at finding niches and defending them. But what of the smallest entities of all—the microcomputers? Rabbits, of course, because they're everywhere and breeding rapidly.

Elephant Walk

As late as 1976, mainframes—the beasts of burden of the electronic world—still commanded 80 percent of the market for all computers. But these ponderous pachyderms got stuck in the mud somewhere along the way and are currently losing ground fast to their smaller and decidedly swifter competitors: the minis and the micros.

The history of mainframe manufacturing, which hit its stride in the 1950s and 1960s, yields few names of individual entrepreneurs who struck it rich. This is because the early producers tended to be well-established public companies with a diversified ownership—such giants as Remington Rand, Sperry (Univac division), RCA, General Electric, and National Cash Register.

The exception was the International Business Machines Corp. (IBM), the successor firm to the Computer-Tabulating-Recording Company (C-T-R). C-T-R owned the patents for one of the earliest punch-card calculating systems designed by Herman Hollerith to tabulate the 1890 United States Census.

Thomas J. Watson, Sr., the business machine salesman par excellence and management martinet, assumed the presidency of C-T-R in 1914 and was a major stockholder by the time IBM started making commercial computers in the early 1950s. As IBM gained supremacy in the computer field, the Watson family likewise flourished. To glimpse how much, imagine your grandfather had purchased a hundred shares of C-T-R stock for $2,750 the year Watson took over. By 1971, those shares were worth $21 million.

When Watson, Sr., died in 1956 at age eighty-two, his financial stake in the company was worth almost $100 million and passed to his wife, two daughters, and two sons (his namesake, Thomas J. Watson, Jr., and the late Arthur K. Watson).

Several other individuals have been winners on a smaller scale in the mainframe computer sweepstakes.

Chicago-born Max Palevsky, with degrees in philosophy and mathematics, aspired to a career as a university professor but found

IBM

The late Thomas J. Watson, Sr., was the computer industry's first multimillionaire. He built IBM into the giant it is today. Here Watson is seated at the console of the IBM 701, the company's first production-line electronic digital computer, designed for scientific work and introduced in 1952.

the ambiance of scholarly life too slow and passive. Then, after a stint at Northrop Aviation in Los Angeles as a staff mathematician, Palevsky helped found a computer division at Packard-Bell.

The entrepreneurial impulse was strong in Palevsky, so strong it got him into trouble. Palevsky was dismissed from his Packard-Bell job for championing the notion of having the parent company spin off its computer subsidiary as a separate company issuing its own stock. But if Packard-Bell wouldn't do it, Palevsky decided he would. He put together a highly motivated group that included Bob Beck, a colleague from Packard-Bell, and Arthur Rock, a young Californian who had just left Hayden, Stone to start his own venture-capital firm. With an initial capitalization of $1 million—$100,000 of it contributed by Palevsky and Beck—Scientific Data Systems was incorporated in 1961 to make small mainframe scientific computers.

Scientific Data Systems prospered. Some 40 percent of SDS's business was tied in with the federal government's freewheeling and well-funded space program. Its earnings trajectory was so dazzling, in fact, that in 1969 its owners sold out to Xerox Corp. for almost $1 billion. Palevsky's share came to $100 million, making him for a number of years Xerox's largest shareholder. Rock's came to about $60 million.

What Palevsky did next was classic. He proceeded to divorce his wife—gossip has it she walked out of the courthouse $35 million richer—and adopt the life-style of a hip capitalist bachelor residing in one of the tonier sections of Los Angeles.

But for Arthur Rock, a sober-minded Harvard M.B.A., SDS was only the beginning. He still plugs into venture-capital deals and always seems to have his name on the ones that net the most money, Apple being just one particularly spectacular example. Others have included Teledyne, Intel, and Diasonics. By 1983, his personal worth was estimated at more than $160 million.

While Palevsky was still laboring at his proletarian job at Northrop in the early 1950s, William C. Norris, another would-be millionaire, was a principal in the small St. Paul firm of Engineering Research Associates. Over Norris's objections, it was acquired by Remington Rand in 1952.

Norris, a spare, laconic electrical engineer hailing from a Nebraska farm, didn't like being someone else's organization man. In league with a number of his associates, he formed Control Data

Corporation in 1957 to manufacture state-of-the-art scientific computers under defense contracts.

To raise capital, the Control Data group did something highly unusual. It sold 615,000 shares directly to the public with neither the financial nor the moral support of an underwriter. Finding buyers proved easy. The entire $1-per-share offering was sold out to friends and former colleagues before the securities were even officially issued. Norris bought seventy-five thousand shares, giving him a 12 percent equity interest, an investment that was worth a high of approximately $55 million in the boom year of 1967. But by 1982, Norris was down to 2.3 percent worth, about $16 million.

Success invariably spawns imitators, often from within one's own ranks. Like IBM, Control Data has had to contend with strong spin-off firms offering a maximum of competition. Control Data's bête noire is Cray Research, Inc., founded in 1972 in a Minneapolis suburb by Seymour R. Cray, Control Data's chief designer.

Cray Research makes the Rolls-Royces of the computer world: multimillion-dollar, high-speed, scientific supercomputers, and it competes head-to-head with its progenitor firm, Control Data. Indeed, Cray's mainframes are so expensive and specialized that it had only fifteen of them in production in 1983.

Cray Research hasn't made Seymour Cray especially rich yet—he owned 2.8 percent of the stock worth about $12 million in 1982—but it's certainly given him satisfaction. While the $3 billion Control Data Corp. sneers at its crosstown rival, pledging to stomp out the "crayfish" and "crawdads," Cray jeers right back since, in ten short years, Cray Research has managed to capture 80 to 90 percent of the narrow market for these highbrow supercomputers. At best, there are about two hundred organizations worldwide—government superagencies, universities, scientific research labs, large aerospace and energy companies—that need these top-of-the-line products. But the market is expanding by 40 percent a year.

Because of the Cray juggernaut, Control Data has diversified into the less prestigious but higher-volume task of manufacturing a broad line of mainframes and related software products in addition to offering extensive data processing services and a line of educational software under the name Plato.

* * *

While Control Data has taken a reactive stance in light of its competition from Cray Research, IBM always manages to stay on the offensive against its dwarf-size rivals, the "plug-compatible manufacturers" or PCMs. PCMs produce mainframe computers or peripheral equipment that run on IBM software. IBM retaliates against the PCMs by keeping them guessing about its marketing and technical strategies. But the best of the PCMs remain undeterred.

Carlton Amdahl, one of the three cofounders of the newcomer Trilogy Ltd., which had expected to market its first IBM-look-alike mainframe in the mid-1980s, said in an interview soon after its 1980 founding: "IBM is wonderful to compete against. They're tough but reasonable. It's a well-run company and, by and large, its salesmen are ethical. They don't cheat."

His words have an ironic ring in light of Trilogy's subsequent setbacks. First the company postponed delivery of its mainframe, a supercomputer to be built around a revolutionary, giant-size silicon wafer that Trilogy engineers were designing in-house. But the company's next announcement several months later was even more startling: It would no longer attempt to produce a mainframe at all but would concentrate instead on supplying other computer manufacturers with its 2½-inch-square silicon wafers. Each wafer would be capable of performing the work of 100 ordinary microchips. Then it announced it was scuttling those plans too. Now it will simply design and assemble circuit packages from conventional semiconductors for computer hardware manufacturers.

Should the company go under, which many analysts predict, it will be a sad day for Trilogy's prime mover, Carlton's father, Dr. Gene M. Amdahl. He's the ex-IBM designer who pioneered the whole PCM movement when he left in 1970 to start Amdahl Corp. This brash little company was the first to challenge IBM to direct product-to-product combat in the mainframe arena.

Amdahl Corp.'s initial machine was larger than IBM's biggest machine, yet it cost far less. But what really piqued IBM was that the Amdahl computer ran on IBM programs and was compatible with IBM support equipment. Sure enough, Amdahl Corp. immediately made inroads with IBM's "loyal" customers, at least with those that wanted to economize.

Gene Amdahl still claims he's counting on Trilogy to push him onto the richest Americans list since his first enterprise didn't get

Gene (seated) and Carlton Amdahl, father and son, teamed up in 1980 to found Trilogy Ltd. to manufacture a look-alike IBM mainframe. For both Amdahls, it was their second start-up.

him anywhere close. That time, with negligible money of his own, he had to exchange most of his equity in Amdahl Corp. for venture capital from Fujitsu, the Japanese computer giant, and from a group of well-heeled American investors, all of whom made out like bandits when the company went public in 1976. All Gene had to show for his effort was less than 2 percent of the stock worth a niggling $2.8 million.

Until Trilogy's design problems surfaced, father and son (their partner, Clifford J. Madden, Amdahl's former chief financial officer, died in 1982) had managed to raise more than $250 million and give away relatively little equity, largely on the strength of their technical reputation. The partners had structured a complicated financing deal that ensured that their percentage of equity would always give them the controlling interest. Now those plans are awry.

Despite Carlton Amdahl's youth—he's in his early thirties—Trilogy is not his first entrepreneurial venture either. As a graduate student at the University of California at Berkeley in 1977, Carlton designed Magnuson Computer Systems' first product, a medium-size IBM-compatible computer, and was recompensed with company stock. But the venture only made him little-league rich. Whether Trilogy will make him any richer is now an open question, especially since he resigned from the company (no reason given) in the spring of 1984.

Monkey Business

In the same way that minicomputers in their heyday in the 1970s edged out many mainframes, microcomputers in the 1980s are now displacing minis.

In 1982, for the first time microcomputers surpassed minicomputers in sales. The reasons weren't hard to grasp. What small-business owner would choose an expensive minicomputer that might require customized software and a data-processing professional to run it when he could get a less expensive, albeit less powerful, microcomputer replete with a wide assortment of cheap, off-the-shelf software? And, most important of all, one that any layperson could run.

In the face of the micro challenge, the predominant minicomputer manufacturers—DEC, Hewlett-Packard, Data General, Management Assistance, Datapoint, Prime Computer, Qantel, and Microdata—are scrambling for new territory to inhabit. Instead of aping each other, most are going their separate ways in search of overlooked niches.

Qantel, for example, is tailoring its minicomputer systems for sports teams. The San Francisco 49ers' head coach plotted the team's 1982 Super Bowl victory with the assistance of Qantel's Sports-Pac program. Likewise, other minimakers are searching for that better idea, the one that will enable them to ward off all comers who threaten their habitat.

Actually, it was a start-up company spun off from Hewlett-Packard in 1974 that introduced the most daring concept in the minicomputer world. The idea was this:

Since so many major industries (the airlines, banking, retailing)

Hewlett-Packard

David Packard (seated) and William R. Hewlett founded the company that bears their names in 1939 in a Palo Alto garage near the campus of Stanford University, where they were both graduate students. Hewlett-Packard Co. was one of the first high-tech firms to settle in the region now known as Silicon Valley. A number of former HP employees—including Jim Treybig (Tandem), Steve Wozniak (Apple), and George Hwang (Integrated Device Technology)—left to start spectacularly successful firms of their own.

relied by the 1970s on computer systems not just for behind-the-scenes accounting and inventory purposes but also to transact business directly with their customers (e.g., make airline reservations, check credit balances), why not design a minicomputer system guaranteed not to fail? Such a system would protect airlines from losing reservations and keep banks from losing funds transfers, thus large sums of interest, when their computers unexpectedly "crashed." Unfortunately, their computers seemed to "crash" regularly, with a minimum of four to five hours a week of "downtime."

Tandem Computers, Inc., founded by James G. Treybig, James Katzman, Michael Green, and John Loustanou in Cupertino, California, offered its first "NonStop" system to the public in 1976. Tandem's NonStop system is really two minicomputers linked together (hence the name "tandem"), each with its own memory. While one computer is handling a transaction, the other computer can be updating the data file. And should one computer fail, the other can take over its duties.

The very existence of such a "fault-tolerant" or "fail-safe" interactive system made such popular services as automatic teller machines possible for the first time. Using these crashproof systems, bank balances could be updated instantaneously instead of once a day, the way it used to be done by mainframe batch processing methods.

Tandem, the pioneer, had the fail-safe computer market all to itself for its first six years of operation, and the company doubled its sales every year. The company also made its four founders comfortably rich, and they're going to be a lot richer if the company shoots up from the second *Fortune* 500 list to the first over the next decade, as most analysts predict.

Besides its product line, Tandem has another thing going for it: its employee morale. Jim Treybig, a transplanted Texan whose twang belies his laid-back California style, says his company represents "the convergence of capitalism and humanism." By that he means the company's site looks more like a country club than an industrial facility, and it boasts one of the lowest employee turnover rates and highest productivity rates in the industry.

Although the company's tennis and volleyball courts, swimming pool, Friday afternoon beer busts, flexible hours, and regular sabbaticals (six weeks every four years) may be responsible for a lot of Tandem's rah-rah spirit, the stock options available to *all* employees, from the lowliest to the most exalted, probably have even more to do with it. Since Tandem went public in 1977, some twenty-five of its mid-thirtyish managers have become millionaires.

Tandem

The four founders gather around a cake to celebrate the delivery of Tandem Computers' one thousandth minicomputer central processing unit in June 1980, just four years after the company began. By 1983, Tandem had shipped over four thousand CPUs. Pictured (left to right) are: John C. Loustanou, Michael D. Green, James A. Katzman, and James G. Treybig.

But by the early 1980s, Tandem found itself sharing the Jungle Jim with those proverbial late arrivals whose antics weren't nearly as original.

Stratus Computer is the first of several entrepreneurial companies to produce a competing fail-safe minicomputer system. Stratus was founded in 1980 by two former Data General Corporation engineers, William E. Foster and Gardner C. Hendrie, and Robert A. Freiburghouse, the founder in the 1970s of a software firm.

Stratus is unique because it's the only one of the start-up fault-tolerant minicomputer firms that lived up to its business plan. It is a textbook case of how to start a company. Stratus shipped its first system exactly twenty-four months after its incorporation. Thirty months after its founding, Stratus went public. The founding trio, each with 4 to 5 percent equity, were suddenly $8 million to $10 million richer.

Stratus, headquartered on the East Coast in Natick, Massachusetts, incurred Tandem's ire when Stratus's early advertisements made explicit comparisons between the two companies' minisystems, implying the Stratus system was superior. Tandem responded with a suit, long since settled, alleging false advertising.

Synapse Computer Corp. was formed within six months of Stratus by two equally eager entrepreneurs, Mark Leslie and Elliot Nestle. One of Synapse's financial backers is Jesse Aweida, the successful computer peripherals entrepreneur who aspires to get even wealthier from venture-capital investments of his own, a pattern that's becoming increasingly familiar within the high-tech world.

Leslie and Nestle bill their $350,000 Synapse system as bigger than the competitors' entries. They say Synapse N + 1 is really a superreliable mainframe with microprocessors providing the redundancy.

Other nascent companies are still struggling to unveil their first system. But it's certain that the founders and financial supporters of Parallel Systems, Inc., of New York; Auragen Systems Corporation of Fort Lee, New Jersey; Sequoia Systems of Marlboro, Massachusetts; and Tolerant Transaction Systems, Inc., of San Jose all have intergalactic personal monetary goals, especially in light of various research firms' forecasts that the fail-tolerant trend soon will engulf the whole minicomputer industry.

Signs certainly point in that direction. While the old-line minicomputer makers—the DECs, Hewlett-Packards, and Data Generals—haven't as yet built minicomputers with redundant architecture, many are beginning to institute programs guaranteeing that their systems will be "up," or fully functional, a set percentage of time in a given period. According to experts, the uptime programs—sporting such names as UpTime Guarantee and WangCare—are mostly public-relations gestures to take the wind out of the sails of Tandem et al.

Rabbit Run

The trouble with rabbits is they tend to overdo a good thing. So, too, with microcomputer manufacturers.

Just as the 1970s incandesced with minicomputers, the 1980s promise to see millions of microcomputers everywhere—on every corporate manager's desk; in the classrooms of every school, college, and university; and in most middle-class American homes.

While that kind of consumer saturation will require the manufacture of several million micros a year, it won't require the full output of the over two hundred micromakers in business in the early 1980s. And all those micromanufacturers, clutching rabbits' feet for good luck, know it. All are hoping they'll be one of the fortunate dozen or so firms that are still in the race at the finish in the 1990s.

Actually, by 1983 the microcomputer market was already beginning to groan under the weight of excess production. Many companies had long since quit because they saw no chance of survival (Technical Design Labs, Processor Technology, Ruben, Solid State Scientific, Findex, Archives), while scores of others had slowed down to a crawl and were soberly reassessing their chances.

Their chances weren't good, for by 1983 any student of industry life cycles could see a familiar pattern coming into focus. Those microcomputer manufacturers already in the race stood a reasonable chance of surviving *provided* they already had a solid market share, sound management, and enough surplus capital to tide them over the rough spots. But few, if any, new entrepreneurial companies stood any chance whatsoever of strong-arming their way onto the racecourse at this late date. The entry fees were far too high.

Microcomputer technology, once so simple that a first-year undergraduate electrical engineer could grasp the concept, had progressed to the state where thirty-two-bit, third-generation microcomputers required $30 million to $50 million in R&D funds for design. In fact, $50 million was the amount Apple Computer spent to develop "Lisa," the name of its goofproof, third-generation supermicro introduced in 1983. Indeed, as early as 1981, Steve Jobs in his speeches was telling competitors that the cost of entering the microcomputer marketplace was "about 30 megabucks at this point in time."

Portables for Pygmies

No one is quite sure who said it first. Maybe it was Jobs of Apple or Chuck Peddle, the premier microcomputer designer, or trade magazine publisher David Ahl. No matter. Whoever it was captured the essence of the next computer revolution when he said, "Never trust a computer you can't pick up."

Dr. Adam Osborne, a suave English expatriate with a nose for giving the public what it can afford rather than what it might dream about, found a way to beat the high cost of entry. A chemical

Osborne Computer

Adam Osborne pioneered the concept of the portable computer. Here he sits beside the Osborne 1, which many said looked like a miniaturized version of a DC-3's instrument panel. Nevertheless, it sold well for a while until a cluster of factors drove Osborne Computer into bankruptcy.

engineer by education, a technical book publisher and columnist by vocation, Osborne decided to shrink the size of computers instead of trying to compete in the glutted second-generation micro marketplace or third-generation R&D labs.

Osborne's gimmick was simple: Design a portable computer that can be trucked around in a suitcase and offer it to the public for the amazingly low price of $1,795 with five well-known software packages thrown in for good measure. Some likened Osborne's marketing strategy to selling the software—purchased separately, the five programs would cost almost as much—and including the computer free.

When in 1980 Osborne announced what he intended to do, the chorus of nay-sayers was deafening. Osborne's computer industry rebellion resulted in a machine that weighed twenty-four pounds, packed as much computing power (though less memory capacity) as the average second-generation micro, had a five-inch diagonal screen, and fit under the seat of an airplane. Presto! The Osborne 1 was one of the world's first *portable* microcomputers—but not the first. That distinction belongs to a fellow Britisher, Clive Sinclair, who introduced in 1980 (a full year before Osborne's entry) a twelve-ounce, book-size microcomputer, albeit with much less power than the Osborne 1. Its memory was only 2K.

Osborne and Sinclair were following the same siren song: Build it simple, build it cheap, and build it in high volume. Osborne's computer was all those things, even if it did look like a miniaturized version of a DC-3's instrument panel. Nonetheless, the world's ugliest computer sold ahead of all expectations—at a rate that placed it among the top three microcomputer producers of 1982, its first full year of production.

At the zenith of his entrepreneurial career, Adam Osborne—purporting to model himself after Henry Ford, Sr.—vowed to continue to build sales by pushing the price down and the volume up. Little did anyone realize how like Henry Ford he would prove to be. Just as Henry Ford's first car enterprise failed within two years, so did Osborne's first computer venture. In September 1983, Osborne Computer Corp. filed for protection under Chapter 11 of the federal bankruptcy statutes. Observers delivering their postmortems said the parable of Adam Osborne was a clear-cut case of entrepreneurial hubris.

Adam Osborne believed he was invincible because his rise had been far too easy. For one thing, journalists loved Osborne, the maverick computermaker. He came from their ranks and knew how to feed their craving for outrageous remarks and unforgettable characters. It's ironic that this flair for generating publicity led to his downfall.

Adam Osborne made the fatal mistake of announcing a new portable computer model—one that was compatible with IBM's ubiquitous microcomputer, no less—months before his chaotic company could deliver it. Naturally, the press cooperated with an avalanche of glowing publicity. But the unexpected result was an avalanche of canceled orders for the company's mainstay, the Osborne 1, and new orders for the stalled "Executive 2." Within the space of one month in the spring of 1983, Osborne Computer Corp. became a company with zero income.

The cash crisis exposed the company's overall management weaknesses. Adam Osborne might be an entrepreneur with all the traits and braggadocio required to transform a concept into a viable product, but he was no professional manager. In two short years his company had grown so fast it was approaching $75 million in gross sales. But during that time, Osborne had failed to institute strong financial or inventory controls or a workable management information system. The company's disruptive move to new quarters in 1982 at the peak of its success didn't help matters, either.

True to form, Adam Osborne walked away from the corporate debris claiming he had no regrets. "You've got to fail some of the time or you aren't trying hard enough," he said. "I don't feel personal embarrassment. You cannot be the kind of person who takes failure hard and be an entrepreneur."

The business world may not have heard the last from Adam Osborne. He admits that being "independently wealthy" is still his goal.

Osborne the fallen pioneer went down leaving an ocean of imitators in his wake. His miniaturization notion had spawned a flood of clones with names such as Kaypro Corporation, Grid Systems Corp., Teleram Communications Corp., Compaq Computer Corp., Otrona Advanced Systems, Seequa Computer Corp., and Gavilan Computer Corp.

Like Osborne, few were given much hope to succeed, with the exception of those *foresighted* entrepreneurs who had designed their machines to be compatible with IBM's microcomputer. Within two years of its debut, IBM's "PC," or *P*ersonal *C*omputer, had become the industry standard and seized the lead in market share from Apple Computer.

Suitcase-size computers may be small, but they're still not small enough to satisfy many people's desire for portable computer power. In the mid-1980s, a portable microminiaturization trend is sweeping through the ranks of the giant computer manufacturers. Companies such as Hewlett-Packard, Radio Shack, and Texas Instruments are doing the entrepreneurial companies, with their suitcase-size micros, one better and producing tiny, pocket-size computers that looked like calculators. In fact, these hand-held computers stand a good chance of rendering calculators obsolete as early as 1986 according to Egil Juliussen of the market research firm Future Computing, Inc.

Simultaneous with this book's publication, all the big-time operators, including the Japanese, are throwing their products into this ring. The only thing holding up the palm-size computer parade is a dearth of software. But as early as 1982, there was already at least one software house staking out that territory.

3 Computer Parts and Peripherals: *The Suppliers*

Andrew S. Grove, one of Intel Corp.'s founding troika, once boasted, "We [the semiconductor industry] are the salvation of the world. If you market the tools for salvation, you have to be incompetent not to make money."

Intel is one semiconductor company that has hauled in truckloads of it since its start-up in 1968. And this phenomenally successful company owes a good deal of its success to Grove and his production know-how. His no-nonsense view of how to manufacture semiconductor components is as down-to-earth as his opinion of his industry is overblown.

As Intel's production czar in the 1970s, Grove claims he took his assembly-line cues from McDonald's Systems, the hamburger chain —in Grove's words, "a maker of medium-technology jelly beans." With his tongue planted firmly in his cheek, this good-humored Hungarian immigrant coined the phrase "high-technology jelly beans" to describe the millions of integrated circuits (ICs) and other semiconductor products that roll off Intel's assembly lines each week. It is Grove's contention that once Intel learns how to make a new "circuit burger" no matter how infinitesimal in size or complex in structure, then it becomes "McIntel's" challenge to produce as many of them as possible in the shortest period of time.

Finding buyers for all these "jelly beans" has seldom been Intel's problem. Since 1968, when Arthur Rock assembled the necessary $5 million in venture capital, the company has gone from zero to $1.1 billion in sales by 1983—and captured spot No. 272 on the *Fortune* 500 list. No, if the company has a problem, it's continuing to match its own record for turning out awesome, breakthrough semiconductor-based products.

Intel's 4004 microprocessor, the four-bit logic IC debuting in 1971, represented the infant company's opening technological salvo. The industry gasped. But before such determined industry leaders as the strapping Texas Instruments could catch their collective breath and react, Intel topped itself two years later with the 8008, the eight-bit microprocessor that's become one of the semiconductor industry's all-time best sellers. It also set the de facto standard for the nascent microprocessor industry, which, by 1975, already included some twenty other companies. To this day, what Intel designs tends to become the industry standard.

The three founders of Intel Corporation, the technological leader in the semiconductor field, gather around a blowup of the 8080 microprocessor. This eight-bit chip has become one of the industry's all-time best sellers. Pictured (left to right) are: Andrew S. Grove, Robert N. Noyce, and Gordon E. Moore.

The microprocessor—the programmable "computer on a chip"—is the tiny marvel that made the microcomputer revolution possible. It is the fingertip-size logic circuit that is now embedded in all manner of consumer and commercial products. Granted, it took several years before the manufacturers of various and sundry products figured out how to use this new form of Lilliputian intelligence, but once they did, it was as if the 8008 series had been programmed to behave like a zillion teeny vacuum cleaners, sucking bright green one-hundred-dollar bills into Intel's coffers. In fact, at last count, this miniature master control chip was estimated to have twenty-five thousand different applications in products ranging from automobiles and airplanes to door locks and electronic games.

While the microprocessor provided Intel with its initial momentum, that product was only the beginning. If a logic IC* is to be used for anything more than the most elementary and repetitive control function, it must be hooked up to other types of integrated circuits —memory circuits, input-output circuits, synchronization, and other types of special-function circuits, for example.

Grove and his partners, Drs. Robert N. Noyce and Gordon E. Moore, soon realized there was a whole constellation of microelectronic products yet to be designed. Adhering to Gordon Moore's dictum that the number of circuits you can etch onto a silicon chip should double every year or so (also known as "Moore's Law"), Intel forged ahead in these other areas with the goal of becoming a full-line silicon supermarket.

At times, Intel's genius for circuit compression even surprised its own founding triumvirate. In 1976, Dr. Noyce asserted in his keynote address to the prestigious Institute of Electrical and Electronic Engineers (IEEE) that there was, indeed, a limit to the number of circuits you could jam onto a chip. No chip would ever be built to equal the complexity of the IBM mainframe, for instance.

Two years later, his own company introduced the 8086 chip, which gave the first indication that, yes, it could be done. Four years

*A logic IC consists of thousands of near-microscopic on-off switches. The tiny circuits linking these switches can be configured in a variety of ways called "gates." Such gates perform various logical operations—e.g., "and," "or," and "not." Hence the name *logic* IC.

later, the titans of the tiny at Intel *proved* it could be done with the debut of something they called the "432 microminiaturized mainframe"—the 432 or "micromainframe" for short.

The micromainframe is one of several radical hardware and software innovations that promise to revolutionize computing over the next decade. The 432 actually consists of three chips so small they can all fit on a fingertip. Combined with its ancillary ICs, the whole system fits into a box the size of the Manhattan telephone directory, even though the 432 possesses the brainpower of computers that even today are the size of refrigerators.

But those hardware companies whose computers are threatened with obsolescence because of the 432 may have more time to maneuver than they think. Intel's mighty midget isn't about to displace anyone or anything for several years.

The what's-it-good-for problem that plagued Intel ten years ago when it dropped the first microprocessor on an unsuspecting world is magnified with the 432. At least the 4004 and 8008 microprocessors were components, no doubt about it. They could be incorporated into the design of other products without changing the basic product too much. But the micromainframe is more than a component—it's a whole subsystem that will force manufacturers to start from scratch and design totally new end products to meet its impressive specifications.

But architectural challenges aren't the only barrier to widespread acceptance. A dearth of software when the 432 was announced is another formidable barrier to its immediate acceptance. There is also the high price. Although the micromainframe itself is priced under $1,000, there are associated chips and a master set of software instructions to purchase, an Intel licensing fee, and a large outlay for the capital equipment required to manufacture any end products incorporating it.

Of course, the stiff costs are understandable in light of Intel's costs to develop it. By 1982, Intel had already spent $30 million to $40 million in R&D funds and said it expected to plow in another $30 million to complete software development.

With sums this large now required to develop state-of-the-art semiconductor products, even a company with Intel's resources must strain to divert the necessary funds into R&D. Fortunately, Intel's capital investment needs for the near future were solved, at least temporarily, in 1983 when IBM purchased a 12 percent interest in the company. Intel happens to be one of IBM's largest outside suppliers of chips. The deal gave Intel $250 million for research.

Clearly, with such astronomical sums needed for research, it will be the rare semiconductor start-up firm in the 1980s that makes its founders as rich as Messrs. Noyce and Moore have become. By 1982, Gordon Moore's 9.8 percent equity gave him a paper worth of more than $100 million, and Robert Noyce's 3.4 percent made him worth more than $35 million. Unfortunately, Andrew Grove's equity was so minuscule that he is often overlooked as one of the original partners.

The semiconductor industry's ability to invent progressively more advanced technologies is one reason why it's been an industry in a continual state of ferment since its beginnings twenty-five years ago. The other factor contributing to its volatility is competition.

In an industry characterized by ever-changing technology, a patent doesn't provide much protection against companies established solely to grind out lower-priced knock-offs. If you're a technological leader like Intel, why bother to prosecute the offenders when you know your own engineers are about to announce a new product with a hundred times the capability of the old? Of course, in the meantime, the copycat companies—and pirates who steal your proprietary products and sell them on the black market—force your own prices down to marginally profitable levels.

One industry trend stands to mitigate some of the cutthroat competition, although it may not make many individuals wealthy on a grand scale.

Now that the designers of televisions, electronic games, toaster ovens, and other consumer products are used to incorporating microchips into their goods, many are demanding that certain types of ICs be tailored to their specific needs. Off-the-shelf ICs aren't specialized enough anymore. Unfortunately, customized microchips can be prohibitively expensive unless they are purchased in volume.

To meet this demand, semiconductor companies are responding in several ways. Some established chipmakers—Intel, Mostek, and American Microsystems, Inc.—have set up "foundries." These are production facilities where the customers' engineers can help design their own IC's using the sophisticated computer-aided-design (CAD) equipment on the premises.

Meanwhile, a subindustry of start-up companies got anywhere from $3 million to $11 million each in venture capital in the early 1980s to make custom and semicustom chips. Among them are Applied Micro Circuits Corp., International Microelectronic Products, LSI Logic Corp., VLSI Technology Inc., and Weitek Corp.

But many industry observers are skeptical about these fledgling companies' chances for survival, let alone their chances for making their founders and backers superrich. They point out that even $10 million is a drop in the bucket and that as early as 1980 it already required closer to $50 million to build and staff a semiconductor factory.

"Besides," said W. J. Sanders III, the founder of the entrenched Advanced Micro Devices and one of the industry's more outspoken members, "when you have so much money chasing so little talent, you get into some areas of excess."

Sanders's remark is as snide as it probably is true.

An industry grows through a process of displacement. Take the semiconductor industry. The Fairchild Semiconductor Co. was the seminal firm of the 1960s, spinning off satellite companies at a rate averaging one a year. The corporate planets in Fairchild's orbit included Rheem (later Raytheon Semiconductor), Signetics Corp., Amelco (which became Teledyne Semiconductor), General Microelectronics (purchased by Philco-Ford), Advanced Micro Devices, and Intel.

But during the 1970s, Intel overshadowed its parent. Intel's stellar performance made it the prime mover of its own solar system. Intel spun off such semiconductor innovators as Zilog, Xicor, and SEEQ Technology.

What's good for an industry as a whole is not necessarily good for the companies within it, however. Spin-off corporate competitors are hardly something progenitor firms in any industry want to crow about. These days they're more likely to take the opposite tack and immediately slap "disloyal" defecting employees with a lawsuit charging theft of trade secrets.

In the early 1980s, there was a spate of these suits in the semiconductor industry. It seemed almost a knee-jerk reaction by established firms, even though the majority of such suits against ex-employees-turned-entrepreneurs are settled long before they come to trial and not necessarily in the former employer's favor.

* * *

Jerry Sanders (né Walter Jeremiah Sanders III) was Fairchild's maverick marketing director and an exceptionally good one, even though he was abruptly fired in 1968 when new management took over. The reason—a rumor he had called on IBM wearing a pink suit. A year later, he started Advanced Micro Devices (AMD). No litigation marred his company's progress even though it engaged in a game one might call "follow the technological leader."

AMD specialized in "second sourcing" Fairchild's and other innovative companies' products. In other words, AMD manufactured and sold copies of standard semiconductor industry products—memories, microprocessors, and logic and linear ICs—and paid the patentholder a licensing fee.

Why would a customer choose to purchase an AMD knock-off over the original? For two reasons:

First, in an industry as cutthroat and mercurial as semiconductors, few customers feel comfortable relying solely on one supplier for the millions of chips they buy each year. AMD traded on this insecurity by stressing reliability. AMD delivered the goods *on time.* By and large, it also delivered *better* goods.

In the 1970s, the industry's quality control—the assurance that the vast majority of the chips shipped to customers actually worked—was lax. Indeed, this lack of attention to detail within the industry was the wedge that both Jerry Sanders and the Japanese used to corner a substantial market share. Japanese and AMD devices were tested *before* they left the factory to make sure they *all* conformed to specifications.

AMD engineers also modified the designs of some of the chips the company licensed, making them slightly better than the originals. The result was a reliable AMD product for which the customer paid more.

In 1969, Sanders recalls, this seemed like the only entrepreneurial option, given his dearth of funding. Even though Sanders had an electrical engineering degree from the University of Illinois, he had moved into sales early in his career. For Sanders the aspiring entrepreneur this had been an unfortunate career move. It seems that management or technological genius of the kind shown by Drs. Noyce and Moore at Fairchild were the only traits that guaranteed venture-capital funding in the late 1960s. Marketing acumen did not.

Advanced Micro Devices

Jerry Sanders was Fairchild Semiconductor's marketing director when he left to start a competing company, Advanced Micro Devices. Sanders is one of the most flamboyant characters in the otherwise staid computer components business.

But Sanders saw the value of superb marketing even if the venture capitalists didn't. What little seed money Sanders did manage to scrape together in the beginning he funneled into a first-rate sales force. And his hotshot sales force was another important reason for the company's eventual success.

Since those early days, Sanders has changed course several times. In 1975, AMD brought out its first major proprietary device, a large-scale integrated computer logic circuit. The chip brought it industry kudos. Two years later, Siemens, the West German electronics giant, purchased 17 percent of AMD for $22.5 million, giving Sanders the bankroll to turn his engineers loose designing further proprietary products.

Because Sanders was tired of AMD's image as a firm that plied the technological backwaters, he declared the 1980s AMD's "excelsior decade," the decade when the company finally breaks into a leadership position in the semiconductor industry. His *modus operandi* is to supply innovative chips to the burgeoning telecommunication business—e.g., sophisticated ICs for telephones as well as for telephone switching systems.

Sanders is a determined man. Already he's been successful in his attempt to reposition his company within the industry. Today, AMD is the industry's fastest-growing company, ranking fifth among all U.S. chip makers. But Sanders is having more trouble transforming his own image.

It's doubtful Sanders, the new semiconductor industry statesman, will ever shake off his old image as Sanders the flamboyant huckster. Earlier quotes attributed to him—"I care only about being rich, and the success of Advanced Micro Devices, in that order" and "Money may not be everything, but it sure beats whatever is second best"—are too fresh in the minds of his Silicon Valley neighbors.

Then there's the matter of Sanders's ostentatious life-style, underwritten by the over $25 million equity he's got in the company.

The silver-haired, moustachioed Sanders spends much of his time in Los Angeles, either in his home in Bel Air or beach house in Malibu. His Rolls-Royce convertible, sleek white Ferrari, Gucci loafers, and diamond pinky ring may brand him as excessively *nouveau* for a serious businessman residing the rest of the time in an elegant apartment in San Francisco; but they are the expected status symbols farther to the south in glitzy L.A.

Indeed, Sanders's personal style prompted one colleague to exclaim, "He reminds me more of a Gay Nineties bartender."

To Sanders, the poor shanty Irish boy from the South Side slums of Chicago, the status symbols are a source of pride, a dream come true. "What's wrong with being materialistic," he asks critics, "as long as the ways of achieving material success are legitimate?"

In contrast to AMD, SEEQ Technology, a 1981 start-up, chose the exact opposite course. While AMD chose to copy the products of its mentor firm, Fairchild Semiconductor, SEEQ chose to become a technological innovator, in the mold of its mentor firm, Intel. From the moment it opened up shop, SEEQ set out to challenge Intel's leadership position by turning out its own breakthrough products in the standard chip segment of the market.

Five ex-Intel employees are members of SEEQ's founding sextet. But they weren't as lucky as Sanders. Before they'd even made the final financial arrangements with their venture-capital backers, Intel had charged them with a litany of legal offensives, among them filching trade secrets, unfair competition, and raiding corporate staff.

Intel claimed that because of the group's abrupt exit in January 1981, Intel suffered delays in product shipments and wanted SEEQ to pay $10 million in damages. SEEQ's countersuit was based on antitrust charges. It claimed Intel was attempting to monopolize the nonvolatile memory chip market, the niche SEEQ sought to enter. ("Nonvolatile" means the memory chip retains information even when it stops receiving electric pulses.)

Despite the legal morass, the premier San Francisco firm of Kleiner, Perkins, Caufield & Byers went ahead with the deal to provide SEEQ with capital of $11 million. Having done that, it marched into the legal fray with a crosscomplaint of its own. Kleiner, Perkins asked for a total of $61 million in damages against Intel and Robert Noyce for the allegedly libelous statements Noyce had made about the venture-capital firm in a *New York Times* interview.

In the interview, Noyce had explained Intel's lawsuit with the comment "We think there's a question of ethics involved here as to whether the venture capitalist should go into an existing company with the technology and try to drag that out, or whether they should wait for the people to come to the venture capitalist and say, 'We want to leave. Would you finance us?' "

According to Kleiner, Perkins's crosscomplaint, Intel and Noyce were accusing the venture-capital firm of something tantamount to alienation of affection, to borrow a phrase from the divorce courts.

The legal skirmishing was concluded in a matter of months in an out-of-court compromise settlement. All parties claimed victory. It was agreed SEEQ hadn't stolen any trade secrets, although SEEQ did have to pay Intel an undisclosed sum for damages caused by the departure of key Intel employees. SEEQ also agreed not to use certain aspects of Intel's proprietary processes for a year.

When the jurisprudential dust had cleared, Roger Bovoroy, Intel's general counsel, declared that an important precedent had been set, particularly concerning venture capitalists. He felt that, henceforth, "if you start negotiating for venture capital for a group that plans to compete with its employer while some members of the group are still employed there, you can expect to be sued and expect to lose."

He's right about one thing: Venture capitalists, with more money to invest today than there are viable business projects to back, are no longer immune from such suits.

Rather than hinder SEEQ, the litigation actually seemed to help. While the suits were receiving extensive coverage in the trade press, SEEQ was raising more money. In addition to Kleiner, Perkins's $11 million, another $17.2 million was put up between April 1981 and August 1982.

But no matter how successful SEEQ is in bringing to market its version of a new memory chip called EEPROM*—developed in conjunction with Texas Instruments, Intel's archrival—it doesn't appear that SEEQ will be contributing any supermillionaires to the year 2000's list of wealthiest Americans. When the firm went public in 1983, only two of the founders, Gordon A. Campbell and Phillip J. Salsbury, had enough equity—2.8 percent each—to be listed in the prospectus as principal shareholders. But the Kleiner, Perkins partnership may contribute some supermillionaires. The prospectus indicated that the venture-capital firm's 23 percent equity had a paper value of $26 million.

The Producers of Peripherals

Peripheral equipment? It's those ancillary electronic machines—such as disk drives, printers, and video displays—that when linked to a central processing unit comprise a complete computer system.

The central processing unit (CPU) is the "brains" of a computer system. The CPU processes the instructions from the programs (or software) and directs traffic to the other parts of the computer sys-

*EEPROM is an acronym for electrically erasable, programmable read-only memory.

tem, telling each peripheral device what to do. The CPU is the brain. Peripherals are the eyes, ears, legs, arms, and hands of the system. They are the devices that help the CPU make contact with the outside world.

The CPU contains memory chips that receive and store information. To expand the CPU's built-in memory, which is limited, it can be hooked up to a peripheral memory storage device. In the 1950s and 1960s, magnetic tape, similar to audio tape, constituted the additional memory medium. Today many tape units have been replaced by disks that look like 8-inch or 5¼-inch phonograph records. They are "played," so to speak, on a disk drive.

Most people in the peripherals industry would name Alan F. Shugart as Mr. Disk Drive himself. After engineering memory devices for IBM and Memorex for two decades, he founded Shugart Associates in 1973 to make a drive to accommodate a newfangled product developed by IBM called a "floppy disk." The term is descriptive in that this Mylar plastic disk—or "diskette," as it's often called—is flexible rather than rigid. But in a struggle reminiscent of Gene Amdahl's experience with Amdahl Corp., Shugart was ousted by Shugart Associates' financiers.

With little money to show for his investment of time and effort in the company that still bears his name (though it's now a division of Xerox), Shugart cast about for another business notion. Shugart was looking for the one big idea that would offer him the opportunity to get stinking rich the second time around since he hadn't yet pulled it off.

The key idea came from Finis F. Conner, a marketing man who had worked at Shugart Associates and who had just been fired from International Memories, Inc., a division of Dorado Systems. (In Conner's own words, he was booted for "seeking too much control.") Conner proposed they form a company to manufacture Winchester drives for 5¼-inch rigid aluminum disks, the hot new technology on the block.

Winchester memory devices represent another technology pioneered by IBM in the early 1970s. IBM's original Winchester prototype was a dual disk drive encasing two rigid metal disks fourteen inches in diameter. But smaller "mini" disks eight inches in diameter were the ones that actually achieved the greatest commercial following—that is, until 1980, when Seagate Technology, Conner and Shugart's company, introduced a disk reduced even further in size.

Winchester disks, with their thin coating of iron oxide, perform better. They can transfer data to the CPU up to twenty times faster than floppy diskettes, and they hold fifteen or twenty times more information than floppies of an equivalent size. But there are disadvantages, too: They're much more expensive and much more environmentally sensitive than their floppy counterparts. A Winchester drive has its rigid disk permanently sealed inside it, and the whole unit is encased in an airtight compartment so no dust can cause the device to suffer a disastrous attack of amnesia. A Winchester disk cannot be removed from the drive, as floppies can. Thus floppies are a far more flexible memory medium for users.

Shugart Associates was already making an 8-inch "mini-Winches-

Seagate Technology

Alan F. Shugart is often referred to as "Mr. Disk Drive" because he founded Shugart Associates, the firm that made one of the first disk drives for floppy diskettes. Ousted from the company by its financiers, Shugart then teamed up with Finis Conner to start Seagate Technology.

Seagate Technology

Finis F. Conner had just been fired from his job when he approached a former colleague named Alan Shugart with an idea for starting a company. The result was Seagate Technology, the firm that is now the leading producer of 5¼-inch Winchester disk drives.

ter" drive by 1979 when Conner and Shugart talked, but the 5¼-inch "micro Winchester" market was wide open. Because a 5¼-inch drive would be smaller, lighter, and cheaper than the bulkier 8-inch, they reasoned that the microcomputermakers would jump at the chance to offer it as part of their desktop computer systems.

Conner and Shugart reasoned correctly. The original equipment manufacturers (OEMs)* leaped at the chance, making Conner and

*Original equipment manufacturers—referred to as "OEMs"—usually manufacture their own computer, then buy compatible peripheral equipment (i.e., disk drive,

Shugart's company, Seagate Technology, the industry leader. Within three years of its founding, Seagate Technology had seized a 65 percent market share.

When Seagate went public in September 1981, each partner retained 10 percent equity. Their individual equity has since dropped a point or two in subsequent secondary offerings, but the value of their holdings has moved up with the market price to as much as $80 million.

No two men enjoy their prosperity more. Conner, a race-car nut, drives around in a Porsche. Shugart simply spends more time and money doing what he's always done—gambling in Las Vegas and fishing off the coast of Santa Cruz from the stern of his power cruiser, the *Junkfish.*

Another man also got several increments richer as a corporate investor in Seagate. C. Norman Dion, who already had 15 percent equity ownership of Dysan Corporation worth some $30 million, was intrigued when Conner and Shugart first approached him about backing their idea. As head of one of Silicon Valley's fastest-growing magnetic disk manufacturers, Dion had good reason to appreciate the opportunities inherent in Conner and Shugart's proposition. After all, if Seagate ever did get the backing to manufacture its machines, it was going to need disks to put in them. If the right deal could be struck, why shouldn't Dysan supply the disks?

At the time of their meeting in 1979, Dysan Corporation had been producing fourteen-inch Winchester disks for several years and was just beginning to make the eight-inch size. After sitting through the Seagate sales pitch, Norm Dion immediately realized the eight-inch size was the wrong size. At least it did not represent the wave of the future.

On November 14, 1979, Norm Dion gave Alan Shugart a check drawn on Dysan Corporation sealing their agreement. The agreement granted Dysan Corporation an equity interest of close to 50 percent in Seagate. It was understood that Dysan would be Seagate's

---

videoscreen, printer) and software from other vendors at wholesale. The OEMs then sell the complete computer system at a marked-up price. Some OEMs, though, are only packagers, buying all the equipment, including the basic computer, from outside sources.

primary supplier of 5¼-inch disks. Moreover, the agreement gave Dysan a royalty-free, nonexclusive, nonassignable, irrevocable worldwide license to make, use, and sell Seagate's eventual product and to incorporate it into any computer system it might want to market in future.

Alan Shugart, the corecipient of Dion's largesse, was more than just ecstatic. He was also relieved. He'd struck a mutually beneficial bargain with a man he could trust. The two men had known each other for years. Alan Shugart had been Norm Dion's boss, first when they'd worked together at IBM, then at Memorex.

Alan Shugart later recounted the events of that fateful day in November 1979. He recalls, "We had always planned to get the lawyers to document the deal, but we never got around to it—until we raised $1 million in venture capital the following June. Venture capitalists don't like [or understand] handshakes."

While Seagate Technology was shipping out its first micro-Winchester drives in July 1980, some two dozen other companies were gearing up to follow suit. Foremost among them was a young entrepreneurial company sporting the name of its Indian founder.

Tandon Corporation was founded in 1976 in the Los Angeles suburb of Chatsworth by a thirty-five-year-old named Sirjang Lal Tandon (nicknamed "Jugi"). The son of an affluent Punjabi attorney, Jugi emigrated from India in 1960. Having earned two master's degrees from American universities—one in engineering, the other in business—Jugi was determined to make a go of some business of his own choosing.

Tandon Corporation actually was his second entrepreneurial venture. The first was an overnight failure. He'd tried breaking into the motor home business during the 1973–74 Arab oil embargo. His sense of timing was better the second time around. Impressed by the emerging but as yet practically nonexistent market for microcomputers, Jugi Tandon took aim at microcomputer peripherals.

Tandon Corporation's first product was a magnetic recording head, a component for floppy disk drives. Jugi gambled he could sell each head for $18 and still make a profit, even though $40 was the going price for the heads at the time.

Tandon's first customer was Magnetic Peripheral, Inc. (MPI), an Oklahoma City-based division of Control Data. Jugi admits a certain amount of subterfuge went into securing that first contract. Before signing the contract, MPI sent emissaries to Chatsworth to check out Tandon's facilities. At that point, Tandon's "facilities" consisted of one rented room, a telephone, and a garage where Jugi's wife, Kamla, helped him do the required soldering, wiring, and assembly work.

"Before the visit," Jugi recalls with a sardonic smile, "I hired a handful of actors to stand around, look busy, and be part of my company." The ruse worked. "Since then, we've shipped some one million magnetic heads to MPI," he boasts.

Tandon Corporation has gone through a multitude of other changes as well. In 1977, when most diskettes were still *single*-sided,

Tandon

Sirjang Lal Tandon's first entrepreneurial attempt went nowhere. His second made him superrich: In 1976 he founded Tandon Corporation in a suburb of Los Angeles to make magnetic recording heads that are components for floppy disk drives.

Tandon patented its technology for a low-cost, *double-*sided recording head, making it possible to "play" both sides of a disk, thus doubling its storage capacity. In 1979 and 1980, Tandon added the end product—disk drives for both floppy and rigid disks—to its list of merchandise offerings.

By 1981 Tandon had blossomed into a full-line manufacturer that makes *all* the component parts for its disk drives in-house. Moreover, while the company's competitors were paying their laborers wages on a par with American-industry standards, Tandon was paying its employees in subsidiaries in Bombay and Singapore wages on a par with Third World standards. Raw materials are likewise cheaper there.

Jugi claims this "vertical integration" and offshore labor force give Tandon a strategic advantage in an industry where most of its competitors are mere assemblers relying on American workers. Tandon can slash prices. His competitors can't.

For Jugi Tandon, all the wheeling and dealing has paid off big. When his company went public in February 1981, Jugi's 17.2 percent of the stock was worth around $30 million. It's since moved up in value to as much as $189 million.

Tandon stock certificates worth more than $10 million are also lining the safe-deposit boxes of Gerald A. Lembas and Leonard B. Lundin, two of the many key employees Jugi recruited from his own former employer, Pertec Computer Corporation. Jugi claims as many as sixty early employees who were compensated with stock and stock options have become millionaires. He also mentions the clutch of venture capitalists (C. Kevin Landry, Jean Deleage, et al.) who prospered handsomely from their Tandon investment.

The disk-drive manufacturing business harbors another immigrant —this one from Palestine—with an equally obvious compulsion to succeed.

Jesse I. Aweida's Storage Technology Corporation, which he and several partners—all ex-IBM employees—founded in 1969, has made its mark producing IBM-compatible peripheral memory storage equipment. Storage Technology's knack for second-guessing "Big Blue," as some call IBM, helped it roar to the head of this cutthroat market niche despite the fact that its main competitors are three of the biggest, best-financed companies in the business. They are IBM, the mother lode herself; Memorex (a division of Burroughs); and Control Data.

Aweida is hooked on challenges, which accounts for his recent about-face. Once the decade-long rush he got from watching his company zoom to number 322 on the 1983 *Fortune* 500 list subsided, Aweida immediately began casting about for some exhilarating new experience. He'd long felt that the PCM marketplace was far too small for a man of his many talents and energies, so he declared that his next coup would be to develop an optical laser memory device to couple with IBM's mainframe computers. The trick is to do it before IBM does. This advanced technology, when it's perfected, will allow for the storage of almost double the amount of information as today's top-of-the-line mainframe magnetic disk drives and will do it at lower cost.

One of the most unexpected things about Storage Technology is the location of its headquarters. It's situated in the Rocky Mountains town of Louisville, Colorado. But apparently even Louisville, Colorado, isn't far enough away to shield Storage Technology from one of Silicon Valley's major problems—the desertion of key employees, either because they want to start their own high-technology company, or because they want to benefit from a better stock-option deal offered by some entrepreneurial competitor. When it happened at Storage Technology, Aweida sued.

The defendant was Ibis Systems, Inc., a new company making high-capacity computer data storage subsystems. Ibis had recruited thirteen employees en masse to assume important posts in its neighboring Boulder, Colorado, R&D facility. Ibis's home base, however, is nowhere near Louisville. It's in the Los Angeles suburb of Westlake Village.

Printing Yen

Although the Japanese have never been too successful selling their computers directly to American consumers, they've been remarkably successful selling to American OEMs the things one puts in and around computers: semiconductor components and peripheral equipment. It's a servants'-entrance approach that the Japanese hope may lead them into the parlor someday.

The strategy is simple. Japanese manufacturers sell peripheral

Storage Technology

Jesse I. Aweida struck it rich producing IBM-compatible peripheral memory equipment. Today this Palestinian immigrant uses his profits from Storage Technology to invest in other people's high-technology start-up ventures.

equipment—printers and video monitors—to American original-equipment manufacturers. The OEMs, in turn, package the equipment into a complete computer system and slap a familiar American brand name on it. In the early 1980s, the Japanese used this approach to commandeer about half of the United States market for the less-expensive dot-matrix printers. They didn't need to make any special effort to sell their video monitors (known also by the technical name of "cathode ray tubes" or CRTs). For many years, Japanese manufacturers have been supplying a good portion of all the videoscreens that accompany American-made CPUs.

Robert A. Kleist—like Jugi Tandon, an alumnus of Pertec Computer Corporation—has the distinction of being the only American entrepreneur to beat the Japanese at their own game. In the last decade, he alone has managed to make big money manufacturing slow- to medium-speed dot-matrix printers.

Printronix, Inc., his company based in Irvine, California, went public in 1979. That and subsequent offerings drove Kleist's personal worth up to some $15 million by 1982. Kleist boosted his bank account several more increments with his venture-capital investment in Seagate Technology, where he sits on the board.

However, letter-quality daisy-wheel printers—higher-quality and higher-priced than the dot-matrix—are still an American preserve at this writing, primarily because many of the patents for these printers are American-held. But they may be destined to go down as America's last hurrah in the printer arena since the Japanese are now showing an interest.

Unlike dot-matrix printers, which form each letter out of a series of tiny dots, letter-quality printers produce type that looks like a good electric typewriter. But electric typewriters and daisy-wheel printers use different technological means to achieve similar ends. Good electric typewriters carry their typeface on a metal device the size and shape of a golf ball. Letter-quality printers, in contrast, have their characters embedded on the spokes of a flat device that looks like a three-inch wheel—or the petals of a daisy, if you will.

These letter-quality—or daisy-wheel—printers usually have the name Diablo or Qume right next to their serial numbers. Throughout the 1970s, these two American companies were practically the only producers of these higher-priced and higher-quality printers.

David Lee, born in China and raised in South America, perfected the delicate-looking device called a daisy wheel as an engineer at Diablo Systems, Inc., in the early 1970s. When Diablo was purchased by Xerox in 1972, Lee got the urge to strike out on his own. A year later, he did—with $800,000 in venture capital from Sutter Hill Ventures.

Along with the financing, Lee got an entrepreneurial partner. Sutter Hill Ventures recruited Robert E. Schroeder, a 1962 graduate of Harvard Business School, to safeguard their investment. Schroeder would provide the strategic thinking and managerial expertise that Lee, the visionary engineer, supposedly lacked.

Qume, their company, did not prosper. In fact, all Lee and Schroeder ever seemed to do was work ten hours a day, seven days a week.

They couldn't seem to get off the roller-coaster ride moving them from one financial crisis to the next. The tension was running particularly high in 1975 because Xerox had instituted a $20 million suit against Qume charging theft of trade secrets. The suit was later settled out of court.

In the mid-1970s, just when the obstacles appeared the most insurmountable, the company finally found its first level stretch of track. At about this time, word-processing systems came into vogue. Soon they were all the rage with large, progressive American companies in the process of automating their secretarial pools. Diablo and Qume letter-quality printers were the obvious choices to couple with dedicated word-processing systems. Later, they were also the best bets to incorporate into systems built around the microcomputer. Qume's product began capturing a growing market share of this exploding business.

But in those days before "venture capital" became the household term it is today, Qume kept running out of money to finance its sudden, extraordinary bursts of growth. Finally, Qume's need for investment capital became too great and the weary founders sold out to International Telephone & Telegraph (ITT) in 1978 for $155 million. Qume's management, especially Lee and Schroeder, shared $50 million. The remainder went to Qume investors and to pay off debt.

After the sale, Schroeder stayed on as a consultant and continued to maintain a residence for his family—his wife, Ann, and four children—in Palo Alto near Qume. But the Schroeders also decided to live a little. With the $7.5 million he realized from the sale, the couple bought a three-story Victorian house on a crest that overlooks California's Napa Valley vineyards in a town called St. Helena. The Schroeders also own an apartment on San Francisco's snobby Russian Hill.

David Lee, in contrast, has hardly raised his nose from the corporate grindstone. He stayed on as one of the company's senior officers.

4 Software: *Its Publishers and Servicers*

Marketing experts have long maintained that *software sells hardware.* In other words, an ample catalog of excellent programs accompanying your machine almost guarantees its success.

Apple Computer learned this lesson well when VisiCalc—a financial-analysis program written to be run on Apple's Model II—burst forth upon an unsuspecting business community in 1979 and suddenly put Apples on executives' desks all across the country.

But, as circumstances continually prove, even marketing shibboleths need periodic revision. By 1983, that old saw about software selling hardware was certainly due back at the garage for some serious adjustment.

Why?

Because only five years after that unexpected synergy was forged between the producer of Apple microcomputers and the publisher of VisiCalc, these two rising-star companies found themselves working at severe cross-purposes.

Two watershed events in 1983 set Apple Computer and VisiCorp on a collision course. That year, Apple Computer introduced its third-generation, thirty-two-bit micro, dubbed "Lisa." It was the first in a new family of Apple computers that includes the "Macintosh," introduced a year later. On the opposite side of the fence was VisiCorp's offering—an advanced microcomputer program called "VisiON."

It was Apple and VisiCorp's pursuit of an identical commercial goal that brought them to an impasse. The goal was to create a thoroughly idiotproof product. Lisa's designers accomplished this goal by hardwiring many basic software functions into their machines' innards. The Visi^{ON} designers, on the other hand, created a freestanding program you insert into a computer's disk drive to get the machine to behave in the same "friendly" way as the Lisa series.

But think of the consequences for software companies should the Lisa series, with its built-in programs, represent the wave of the future. Clearly, those consequences extend far beyond the fate of one hardware and one software company. Indeed, the Apple-VisiCorp confrontation is forcing the microcomputer world's marketing mavens to rethink the whole relationship between software and hardware.

Personal Software, Inc. (VisiCorp's original name), was founded in 1977 with $500 from Daniel H. Fylstra's student savings account.

That Fylstra—with this mild-mannered, Clark Kent demeanor—would one day team up with other hackers to start a software company was predictable. He claims that as a youngster growing up in San Diego, he first dreamed about starting his own company as a tyke of six or seven, and actually began tinkering with computers at twelve.

The second child of a Protestant minister, Fylstra went on to MIT and graduated with a degree in electrical engineering and computer science. As a lark, he took a part-time job as an editorial employee at *Byte* during this well-known computer magazine's formative years. To earn serious money, he worked as an engineer for Intermetrics, Inc., a Cambridge firm under contract to design software for NASA's space shuttle and for the European Space Agency.

It was an instructive experience for Fylstra. Just one year of exposure to large-scale bureaucracy convinced him that he'd better find out how to run a business of his own, since wending his way up the organizational ladder the conventional way didn't appeal to him. So off he went to Harvard Business School in 1976 to plot his entrepreneurial escape.

A confluence of factors led Fylstra to software publishing. Fylstra's tenure at *Byte* had piqued his interest in the commercial side of the emerging microcomputer phenomenon. At *Byte* he saw that people were, indeed, making a living selling programs for those small computers that only yesterday were nothing more than hobbyists' toys. What's more, these budding tycoonlets seemed to be making money the easy way—through coupon advertisements in magazines like *Byte.* Actually, in those days software authors had no other choice, since computer stores were still few and very far between.

Fylstra also noted that the handful of microcomputer software companies in existence in the mid-1970s comprised a cottage industry in the strictest sense of the word. Most were run by amateurs with no head for business. And they were shaped in the classic technicians' mold: They were long on programming skill and short on marketing savvy.

Jacqueline Thompson

Daniel H. Fylstra was in his second year at Harvard Business School when he and a Canadian software entrepreneur joined forces to start the firm that became VisiCorp. The pair signed up two ace programmers to write VisiCalc, the phenomenally popular financial program credited with making the Apple II the first microcomputer to gain widespread acceptance among young executives.

Fylstra, who thinks in terms of business case studies and certainly talks that way, decided to reverse that equation. Wouldn't it be fun, he thought, to run a software firm whose professional advertising and marketing expertise set it apart from the usual moonlighting programmers who peddled their wares in hit-or-miss fashion?

Personal Software was actually established—as an unincorporated entity—in 1977 as a fun-and-games project for Fylstra's Harvard marketing course. His objective was to mass-market a piece of microcomputer software like you might market a book. With a game program he'd written, that's just what he did.

Fylstra's $500 paid for a direct-mail magazine ad that brought in more in revenues than the cost of the space and thus subsidized further advertising. This self-financing pattern is typical of all the successful microcomputer software firms founded around this time.

That same year Fylstra met his future partner. Peter Jennings, a Canadian with an M.B.A. and a master's in physics, ran his own software operation in Toronto. Jennings had two motives for teaming up with Fylstra: He had a chess program for hobbyists he wanted to market in the United States, and he wanted to emigrate. Jennings ended up following his chess program across the border and subsequently applied for U.S. citizenship.

The pair incorporated Personal Software in 1978 and, again, Fylstra sold the new program through computer magazine ads, specifying COD payment. MicroChess was an immediate hit and grossed $1 million in its first year.

But despite its success, Fylstra sensed that the big money lay elsewhere and that he could get access to this unmined gold deposit by developing software products aimed at business professionals. So the duo deferred any personal profit-taking and poured the money from MicroChess into the business-oriented program that was to put the company—and Apple Computer—on the map.

As Fylstra tells it, Daniel Bricklin—who was one year behind him at Harvard Business School—had the idea for a program that would, in effect, turn a microcomputer screen into an electronic spreadsheet. With such a program, the microcomputer user could type numbers and formulas onto the screen's rows and columns and then do "what if" forecasting. Since the numbers on the screen were related, the computer would do the necessary string of calculations every time one number was changed.

These were just the kind of calculations—heretofore done laboriously with a pocket calculator—that made Bricklin's and his fellow business school students' homework assignments such a chore.

A Harvard finance professor was responsible for the fateful meeting between Bricklin and Fylstra. This professor had earlier told Fylstra he'd never get rich peddling software for microcomputers. Now this same professor was telling Bricklin his spreadsheet notion would never fly either. The professor introduced the two ambitious young men for only one reason: He wanted Fylstra to confirm his low opinion of Bricklin's spreadsheet concept. Fylstra did no such thing.

Flush with the MicroChess profits, Fylstra was on the lookout for new programs to sell. He loved Bricklin's idea and invited him to develop a prototype on the Apple II in Personal Software's third-floor office behind the Business School. Bricklin finished the prototype in three weeks, and Fylstra and Jennings offered him a contract much like a standard book contract: He would get an advance against royalties.

The contract was eventually signed by a corporate entity called Software Arts, a company Bricklin had formed with his buddy Robert Frankston—who happened to be a nonpareil programmer. (Fylstra, Bricklin, and Frankston are all MIT alumni.) Software Arts would earn 35.7 percent of the program's wholesale price and 50 percent on certain bulk orders. The arrangement was generous compared to the standard royalties offered later, when the industry matured.

The program that emerged from the agreement one year and $100,000 (in development money) later was VisiCalc. This applications software package has dwarfed the sales of all other quantitative microcomputer programs before or since. *Datamation* magazine in its September 1981 edition went so far as to hail VisiCalc, by then two years old, as perhaps the most important computer development since the creation of the FORTRAN language "because it sells otherwise unapproachable computers to people without any technical training."

Fylstra takes the credit for coining the name "VisiCalc," code for "visible calculator." Indeed, VisiCalc was so popular and the name so catchy that it also came to serve as the umbrella for Personal Software's family of follow-up products—VisiPlot, VisiTrend/Plot, VisiDex, VisiTerm, VisiFile, VisiSchedule, and, of course, VisiON.

Jim Raycroft

Robert M. Frankston (standing) and Daniel Bricklin are the two programmers who contracted with VisiCorp in 1978 to write VisiCalc. Their Wellesley, Massachusetts, firm, Software Arts, now markets its own line of microsoftware packages.

In early 1982, Personal Software finally went whole hog and changed its corporate name to VisiCorp to foster even closer brand identification with its VisiSeries. By this time the company had also changed its venue to Sunnyvale in the heart of California's Silicon Valley.

Quality products alone fail to account for VisiCorp's early success. Distribution and merchandising were also essential. Fylstra was one of the first microsoftware publishers to sell his products in computer stores, the retail outlets that began appearing in the late 1970s. From the outset he insisted on attractive, distinctive packaging to make sure VisiCorp's products captured browsing consumers' attention. He even offered retailers a stand-alone rack on which to display VisiCorp's merchandise.

Such adroit packaging and careful orchestration of all the available retail distribution channels gave VisiCorp an early edge over its less professional competitors. But in the intervening years, Visi-

Corp's formidable lead has been narrowed almost to nonexistence by a crowd of late-blooming competitors. Many have simply copied Fylstra's sophisticated moves in the marketplace, particularly his brand-recognition marketing strategy.

For example, Information Unlimited Software, a firm located just north of San Francisco in picturesque Sausalito and headed by William Baker, an ingenuous Hoosier in his twenties, has expropriated the adjective "easy" for its EasyWriter, EasySpeller, and EasyFiler programs. A few miles farther north, in San Rafael, Seymour Rubinstein's MicroPro International is the proud publisher of WordStar, America's most popular microcomputer word-processing program, along with a host of helpmate programs—SpellStar, DataStar, and CalcStar. Then there's Perfect Software with its Perfect series, Sorcim with its Super series, and Software Publishing with its PFS series.

On the East Coast, there's the Atlanta-based Peachtree Software, Inc. Peachtree calls its programs "Peachware." In its advertising campaign it exhorts software consumers to "Pick a pak [sic] of Peachware"—such as PeachText, PeachCalc, PeachTax, and PeachPay.

By the early 1980s, packaged software for microcomputers was already big business. In fact, the industry had grown so vast and diverse that for every one of the estimated 200 microcomputer hardware companies, there were twenty to thirty software purveyors—many of them one-man bands—turning out some five hundred new titles a month.

Unfortunately, for the majority of these "publishing houses" the rewards were meager and were in the psychic rather than the monetary realm. In contrast, for the industry's handful of leaders, money was seldom a problem. The cash required to expand those firms emanated from an entirely unexpected source.

During the heyday of microsoftware-company start-ups in the late 1970s, it was understood that venture-capital funds were off-limits to these inexperienced entrepreneurs. Even venture capitalists inured to risk viewed such software investments as suicidal. After all, the best of the microsoftware companies were run by a bunch of flaky, long-haired kids fresh out of college, if they'd even made it *that* far.

When aggregate annual sales of packaged software for microcomputers eventually passed the $1 billion mark, though, these same venture capitalists blinked sleep out of their eyes. Their interest was piqued further when observers began forecasting revenues of $25 billion by 1990.

But the set of cold, hard facts that actually got venture capitalists to reach deep into their pockets was even more compelling: In comparison to hardware producers, software companies require significantly less development capital; their profit margins are much higher, often double; and they're positioned better to compete against their Japanese counterparts.

Moreover, as luck would have it, microcomputer software companies were still virgin territory when the more alert venture capitalists woke up in 1980 to the opportunities. They didn't waste any time. Two years later, nine of the country's better-managed software companies, including VisiCorp, had accepted venture-capital backing.

A number of others demurred. First, these prosperous young software firms didn't need other people's money. They financed their fast growth out of retained earnings. Second, they wanted to share their equity with as few people as possible.

Because of the stampede mentality among the investors, the software companies who did sign agreements practically dictated the terms and, in general, gave up relatively little equity. However, the venture capitalists usually got their way on an issue dear to their balance sheets: They, not the entrepreneurs, could decide when the company would go public—in short, when they'd cash in on their investment.

Fylstra claims his primary motivation for negotiating VisiCorp's multimillion-dollar deal was to attract experienced men like Arthur Rock to his board—men whose managerial advice is known to grow seedling companies into giant oaks. Fylstra won't say how much equity he had to surrender to his seven venture backers and other institutional investors in exchange for their $8.8 million.

". . . Let's just say I'm the largest single stockholder among the two hundred or so stockholders, which includes many of our employees," he told me. Peter Jennings also remains a significant shareholder.

Fylstra is a young man who knows where he and his company are trying to get, even if some recent setbacks do not augur well for VisiCorp's future.

"All my energy is focused on making VisiCorp the country's largest independent software firm. The success of the company as measured by its revenues and my net worth is, after all, just a yardstick. If I doubled my net worth right now, it would be a huge amount, but what would it mean to me? I'm much more achievement-motivated than financially motivated."

There's no doubt that to date VisiON is VisiCorp's greatest *technical* achievement, though hardly its greatest *marketing* achievement. That this technological coup should conflict with Apple's is as unfortunate as it is ironic.

In practice, the similarities between Apple's Lisa (a third-generation microcomputer) and VisiON (a second-generation microprogram) are uncanny.*

Both Lisa—and its lower-priced counterpart, the Macintosh—and VisiON allow users to bypass the computer keyboard when implementing commands. This user-friendly feature is a great relief to the millions of people who don't know how to touch-type. Instead, users manipulate a "mouse," a device similar to the joy sticks and "trakballs" that control video games. The mouse activates an arrow on the videoscreen. With the arrow, users indicate which task they want the computer to execute.

In this simple manner, not only do users avoid the codes that made operating a second-generation microcomputer something akin to piloting a spaceship, but also the new products almost make literacy unnecessary, since many of the commands are depicted by symbols. For example, to call up a mailing list program, all a user might have to do is aim his mouse at the stamped-and-addressed envelope pictured on the monitor in front of him.

Lisa and VisiON's ability to swap information quickly and easily between heretofore disparate programs signals another gigantic leap

*The generations of micro hardware and software are out of sync: The first generation of microcomputers that appeared around 1974 had no software. The hobbyists who bought these computers in kit form not only assembled them but also wrote their own amateur programs.

VisiCalc, for example, is a first-generation program that was written for the Apple II, a second-generation machine.

forward in microcomputer technology. It is also the reason why all second-generation microsoftware products such as VisiON are referred to, generically, as "environment" or "integrated software." As the term implies, integrated programs interact with each other. In Lisa's case, many of its integrated software functions are built right into the machine—hence the acronym "Lisa," which stands for "*l*ocal *i*ntegrated *s*oftware *a*rchitecture."

In contrast, the first-generation software packages are termed "stand-alone programs." It's an apt term because it means just what it says. With stand-alone programs, users have to change disks every time they want to perform a different task. Each stand-alone software program has a separate and distinct function that cannot be merged with any other program's function. Nor can the text or numbers from one program be traded for the material in any other program. A word-processing program cannot be combined with, let's say, a financial modeling program.

Using integrated software, though, a user can create a complete financial report, for example, without once changing disks. To do this he or she might first call up figures stored in a data-base file, change those numbers using a "spreadsheet" program like VisiCalc, insert the results in a bar, line, or pie chart created by a graphics/plot program, and finally include the chart in a report the user writes using a word-processing program. With Lisa and Macintosh, a user can even use the mouse to create freehand drawings and have them inserted in the report if he or she wishes. Perhaps the most miraculous of all, the user can perform these multifarious tasks using commands common to all of them.

Resorting to a metaphor to explain integrated software, you could say that the old first-generation programs are reminiscent of a collection of scratchy, monophonic 78-rpm records. In contrast, the new second-generation programs come already assembled on one neat, digital 33-rpm stereophonic record.

VisiCorp is far from the only—or first—microsoftware firm moving away from the outmoded collection-of-single-disks approach to the more efficient album approach. The first *integrated* program, called MBA, was offered by Context Management Systems of Torrance, California, a privately held firm founded in 1980 by Gib Hoxie and Brian Fischer, two ex-Booz•Allen consultants. Others selling all-in-one software packages are Ashton-Tate of Culver City, California (Framework); Business Solutions, Inc., of Kings Park, New York (JACK programs); Ovation Technologies of Norwood, Massachusetts (Ovation); and Microsoft Corporation of Bellevue, Washington (Windows).

In fact, the integrated program that's made the biggest splash so far remains something called "1-2-3" (as in the phrase "as easy as one, two, three"). It's hardly surprising, since the human brain behind 1-2-3 also wrote, on a free-lance basis, the two statistical programs named "VisiTrend" and "VisiPlot" that are currently part of VisiCorp's family of programs.

Until 1978, when he traded in his stereo system for an Apple computer, Mitchell D. Kapor's mind was still cluttered with counterculture slogans left over from the psychedelic sixties. In short, he was a young man about to turn thirty who'd spent the past decade drifting in search of an occupation. He'd tried several. Suddenly, like many of his chronological peers—now all overaged, overeducated ex-flower children—Mitch discovered the microcomputer. To Mitch and millions like him, the microcomputer represents the new-age recreational replacement for rock music and hard drugs.

Mitch Kapor also discovered he had a facility for programming. And he discovered, to his delight and amazement, that people like Fylstra would pay thousands of dollars for superior software packages. Knowing this, Kapor incurred $30,000 in debts writing the VisiCorp programs, but he eventually recouped that investment fifty times over. He sold the rights to VisiCorp for $1.2 million and quickly earned $500,000 more in royalties.

Then Mitch Kapor got smart. He decided that the average free-lance programmer had a very slim chance of becoming what Damon Runyon called a "rich millionaire" from the 15 to 25 percent royalty fees (based on the package's wholesale price) being offered by the best of the microsoftware publishers. This was certainly true if one published under the meager 4 to 10 percent royalty arrangement offered by the software publishers at the other end of the spectrum. And there were a lot more stingy publishers than there were the share-the-wealth variety.

Thus, Kapor reasoned, the way to clean up was to set up his own software publishing house and become one of the exploiters of free-lance programmers instead of remaining forever an exploitee.

With his profits from the VisiCorp programs, he founded Lotus Development Corporation in Cambridge, Massachusetts, to develop and market 1-2-3. In short order, Lotus attracted $2 million in venture capital from Sevin Rosen Partners, a newly established fund headed by two savvy partners: Benjamin Rosen, the microcomputer industry's foremost analyst; and L. J. Sevin, the founder of Mostek, a respected Texas semiconductor firm since acquired by United Technologies.

Kapor and his associates wrote the initial version of 1-2-3 to run on the one-year-old IBM personal computer and offered the program to the world at a trade show in November 1982. Kapor came away with a $1 million fistful of advance orders even though 1-2-3 wasn't slated for release until early 1983. In an excess of enthusiasm, several industry observers at that time hailed 1-2-3 as the program that would do for the IBM microcomputer what VisiCalc had done for the Apple II.

Lotus Development

Mitchell D. Kapor (seated) was an overeducated ex-flower child until he discovered his extraordinary facility for programming and traded in his stereo for an Apple computer. Here he is pictured with Benjamin Rosen, the well-known venture capitalist who backed Kapor's publishing firm, Lotus Development Corporation.

In fact, just the reverse happened. IBM's micro took off, ringing up sales beyond even IBM's expectations. The success of the IBM Personal Computer guaranteed the success of 1-2-3 because, for many months, it was one of the few integrated programs available for that machine.

By mid-1983, 1-2-3 had displaced VisiCalc as the most popular business program for personal computers. Lotus Development Corporation's venture-capital backers wasted no time turning that fact into crisp, green cash. Lotus stock went on the block for the first time in October 1983. The Sevin Rosen investment group's equity was worth $63 million at the opening price of $18 a share. Mitch Kapor's equity was worth slightly less: $56 million.

Fylstra and Kapor are hardly the first Americans to get rich supplying the instructions that tell computers what to do. In fact, the first software entrepreneurs to enter the ranks of the superrich in the 1960s didn't write programs at all. For good reason. At the time, there was no future in it.

Strange as it may seem today, the earliest computers—those noisy monsters that filled up rooms the size of gymnasiums—didn't require software. Instead, 1940s-vintage computers had their instructions hard-wired into their guts. To change a program, part of the machine literally had to be rewired. It was an inflexible, inefficient way to get any work out of a computer.

Then in a series of scholarly papers—the first appearing in 1946 and entitled "Preliminary Discussion of the Logical Design of an Electronic Computing Instrument"—Hungarian émigré mathematician John Von Neumann and his colleagues proposed the idea of the stored program. The following year, three Bell Labs scientists invented the transistor, that small miracle that would eventually replace the vacuum tube.

These two advances were an enormous boon to computer technology and, together, set the stage for the commercialization of the general-purpose computer as we know it today. Those lumbering "electronic brains" of the 1940s, which were little more than heavy-duty, plodding calculators, were transformed in the 1950s into full-scale information handlers capable for the first time of processing both numbers and words.

The first mainframe manufacturers supplied the software for their machines as part of the system's total cost. Since computers couldn't run without these sets of instructions, it simply never occurred to these manufacturers that programming was other than their responsibility. Besides, convincing their early customers—major American corporations, universities, the military, and government agencies—to buy these arcane, multimillion-dollar systems was hard enough without expecting them to employ an army of in-house programmers as well. In those years, good programmers were scarce indeed. Mainframe manufacturers had little interest in sharing what few programmers there were with their nervous customers.

While it was clear that independent publishers of mainframe software stood little chance of competing successfully against the hardware manufacturers by supplying off-the-shelf programs, there was plenty of opportunity for them in other areas—as software services consultants, for example.

This was especially true after the 1956 consent decree IBM signed with the U.S. Justice Department. IBM agreed to spin off its data processing service bureau into an arm's-length subsidiary called the Service Bureau Corporation. That subsidiary, from the moment of its inception, could no longer enjoy any special advantages from its powerful parent.

H. Ross Perot, hailing from Texarkana, Texas, was one of the first—and now the richest—of those pioneering entrepreneurs who left their well-paying jobs with companies like IBM to strike out on their own as data processing entrepreneurs.

In the late 1950s, Perot worked for IBM in Dallas. But after five years, Perot the supersalesman was bored stiff. The year was 1962. Perot was thirty-two years old and he'd managed to sell his annual quota of mainframes in just three weeks. A young man of action, Perot was not about to sit around for the rest of the year staring at the THINK signs plastered all over IBM's offices. So he quit. And on June 27, 1962, he founded Electronic Data Systems, Inc. (EDS), naming his wife, mother, and sister its charter directors.

EDS would provide broad-based computer services. It would design, install, and operate information systems for any company or large organization that lacked the manpower and resources—or confidence—to manage a facility of its own.

There were plenty of prospective customers around for a lean, hungry firm like EDS, even though, from the beginning, EDS faced a growing cadre of competitors. Still, the data processing (DP) field in general was growing much faster.

In 1966, the five-year-old Association of Data Processing Service Organizations took its first yearly snapshot of its potential constituency and came up with a count of seven hundred data processing firms generating revenues totaling more than $500 million. That figure represented sizable progress from the meager $15 million in collective revenues generated eleven years earlier. But it was nothing compared to the industry's growth over the next fifteen years.

By 1980, the association reported that the computer services field encompassed over four thousand companies and hauled in $15 billion in aggregate annual revenues. The majority of these four thousand companies were small outfits servicing local clients. But EDS was not one of them. Rather, EDS had expanded to the point where it ranked fifth in the lineup of the country's largest computer services firms. Ahead of EDS were the giants in the field: IBM, Control Data, Computer Sciences Corp., and Automatic Data Processing.

But to wealth watchers, the enormous increase in Perot's personal net worth in twenty-two short years is the most impressive statistic of all. When EDS went public in 1968, Perot managed to retain more than 45 percent equity in his firm. Perot stepped up to the cashier's window a second time in 1984 when General Motors Corporation bought EDS for $2.6 billion. Perot's share came to $1 billion.

Perot may be the richest entrepreneur to emerge from the computer services industry, but he's by no means its only self-made multimillionaire.

One of the very first independent information processing firms is the fourth-ranked Automatic Data Processing (ADP). Founded in 1949 when the average American had never heard of computers and had certainly never seen one, ADP made its first buck specializing in the "batch processing" of small companies' payrolls and accounts payable/receivable. ADP employees would fan out over the landscape, picking up and delivering batches of data processed in ADP's headquarters in Clifton, New Jersey.

But that method of servicing customers has long since withered into obsolescence. The new automated telecommunications technology is now making it possible for service bureaus like ADP to transmit data to and from their clients' offices over long-distance telephone lines.

ADP founder Henry Traub, a Polish immigrant, and Frank Lautenberg, an early employee who was chairman until his 1982 election to the United States Senate, are ADP's principal shareholders. They'd like nothing better than to see ADP increase their equity over its current market value of $10 million to $20 million each. They see this happening as ADP gradually converts its eighty thousand customers—and attracts new ones—using as the bait the more modern method of delivery termed "remote" or "online" data services.

As IBM goes, so goes the industry. That maxim was never truer than in 1969. In that year IBM, because of a press of private lawsuits and an impending government antitrust suit, was forced to "unbundle." Henceforth, IBM would no longer offer its customers the "total systems solution." Before unbundling, the cost of IBM's software, training courses, and systems engineering services were included in the price of the hardware. Unbundling meant that IBM would now charge for each separately.

Big Blue had resisted unbundling as long as possible. Just as the 1956 consent decree had done in the case of data processing services, IBM's management knew that this radical change in pricing policy would open gigantic cracks in its once ironclad relationships with its customers.

IBM's management was right. Unbundling created whole new subindustries. Where once there was little or nothing, in the future there would be a host of entrepreneurial service companies specializing in everything from DP training and custom contract programming to hardware repair. And inevitably there would be lots of publishers selling standard software packages for mainframes. There might even be new companies offering a mixture of these products and services.

Management Science America, Inc. (MSA) proclaims itself "the country's largest independent supplier of applications software." It wasn't always so.

MSA began life in Atlanta in 1963. It specialized in computer control systems for the myriad textile and garment manufacturers in the region. MSA gradually branched out into other custom programming areas until by 1970, a combination of bad management and increased competition resulted in a loss of $7 million net on sales of only $9 million. MSA's lenders were horrified and brought in a young turnaround manager with knowledge of MSA's problems.

John P. Imlay, Jr., arrived bearing a hatchet. Imlay's first official act was to reorganize the company under Chapter 10 of the federal bankruptcy statutes. His second was to ax more than seven hundred employees, including the man who had previously hired and fired him twice in the late 1960s from MSA's vice-president-marketing post. Why had Imlay been fired? For insubordination. In true prophet-of-doom fashion, Imlay had repeatedly offered the president unsolicited advice about how to reform the company's free-spending ways.

In the decade since Imlay's purge, MSA has grown progressively stronger and larger. Imlay and William M. Graves, the only founder to survive the bloodletting, steered it into the segments of the software industry earmarked for the greatest expansion. In the wake of IBM's unbundling, for example, MSA was one of a small group of established companies poised to rush into the software void—which it did. MSA's programmers were diverted from custom work to develop standard software packages for mainframes, mostly in the areas of financial, human resources, and manufacturing applications.

But lately, Imlay's strategy is something else: expansion and diversification through acquisition. In June 1981 MSA acquired a neighbor, the two-year-old Peachtree Software, securing for MSA a well-positioned beachhead in the burgeoning microcomputer software marketplace. A year later, Imlay grafted onto MSA a specialized mainframe software company called Arista Manufacturing Systems, a former division of Xerox.

MSA's strategy duplicates the overall consolidation trend in the industry. It's an obvious course to follow in an industry that's still young and highly fragmented. In fact, the industry is so fragmented that in 1981 it was estimated there were forty-five hundred computer software/services companies vying for a market in which no single competitor had more than a 5 percent share. Furthermore, there was —and still is—a drastic shortage of programmers for those forty-five hundred companies to hire.

It's hardly surprising that throughout the early 1980s, W. T. Grimm, the prestigious Chicago research firm, consistently ranked computer software/services as one of the country's most acquisition-prone industries. Indeed, this heady mergers-and-acquisition growth pattern of companies like MSA is characteristic of the industry.

Over the past decade, Imlay has been given plenty of incentive to rise from the ashes of MSA's immolation. In fact, by the time MSA went public in 1981, Imlay had become the company's largest stockholder, with more than 25 percent equity worth about $30 million. In comparison, Graves's stock was worth only half as much. Eugene W. Kelly, who signed on around the same time as Imlay, was the second largest stockholder, with a $15 million stake.

Cullinet Software, Inc., John J. Cullinane's corporate baby, has several firsts to its credit: After IBM unbundled, it was the first software publisher to offer an IBM-compatible report generator, and it marketed the first integrated data-base system, again for IBM machines.

Today Cullinet is situated on Route 128, the famous high-tech boulevard that rings Boston. But its corporate digs weren't as centrally located nor as sumptuously appointed back in 1968, when J. J. Cullinane founded his firm.

In 1970, Cullinet's first product was available. It was that IBM-compatible report generator, called CULPRIT. Other related products followed, but it wasn't until 1973 that Cullinane got his lucky break.

In retrospect, Cullinane admits that the "lucky break" was a long shot. It was an unusual proposition, to say the least: The management of B. F. Goodrich, the tire manufacturer, asked Cullinane if he would like to market a data-base management package it had developed for use on its own IBM mainframe. They claimed their systems software made an IBM computer function like a superb reference librarian—sorting, organizing, and indexing very large banks of computerized files. The program was so efficient that they thought Goodrich should make some money from it. In short, they thought it deserved to be "published."

Cullinane agonized. The program was twenty times the size of his existing software products, would command a huge price, and might sink his business if it developed too many bugs. He took a deep breath—and plunged. To his relief, the program—dubbed Integrated Database Management System (IDMS)—was virtually error-free and a huge success with the owners of IBM mainframes.

IDMS is the product around which J. J. Cullinane built his company. And IBM medium-to-large-scale computers continued to be the only hardware on which the company's products functioned until 1982, when J. J. Cullinane began a headlong rush to diversify. He purchased Computer Pictures Corporation to facilitate Cullinet's entry into the microcomputer software market and negotiated joint ventures with a half dozen other applications software houses. In 1983 he introduced a set of networking programs that enable IBM's personal computers to retrieve data from IBM mainframes. To come are similar programs giving Apple's Lisa the same capability.

When Cullinet floated its initial stock offering in 1978, the investing public learned that John Cullinane held almost 15 percent equity, worth about $3.3 million. His family controlled another 11 percent, worth about $2.7 million. At the time, the total shares outstanding were worth only $24.6 million, but in the intervening years that figure has been magnified severalfold by enthusiastic investors.

But back in the late 1960s, Cullinane wasn't the only Bostonian with the cagey notion that unbundling would create a bustling market for standardized programs for mainframes. After watching several of their co-workers leave IBM to found moderately successful software businesses, James McCormack and Frank Dodge left their respective marketing and engineering jobs in 1969 to do the same.

Jim McCormack was actually a trained accountant with bachelor's and master's degrees in the field and experience working for General Motors and the Big Eight accounting firm of Coopers & Lybrand. Thus it was hardly surprising when McCormack & Dodge Corporation's first software product was a fixed-asset accounting system to run on an IBM mainframe.

When Dun & Bradstreet acquired the firm in 1983 for more than $50 million, Jim McCormack and Frank Dodge were the major stockholders. They agreed to continue at the helm for another three years and will get an additional payout if the company holds on to its position as one of the world's foremost producers of financial and accounting software for mainframes and minicomputers.

Sandra Kurtzig may be one of the few young women to found and manage a software firm. And she's certainly the only woman who can claim responsibility for nurturing a spectacularly successful one. But it's more than her gender that sets her apart from her competitors.

When IBM "unbundled" in 1969, Frank Dodge (seated) and James McCormack were two of the many entrepreneurs who jumped into the void with software programs for mainframes. They are shown examining plans for a new office building for their company, McCormack & Dodge, in Needham Heights, Massachusetts.

Kurtzig has managed to take a company she started with $2,000 in savings in 1972 in her family's second bedroom and shepherd it to the point where its stock could be offered to the public nine years later—and after that offering, she still remained by far the company's largest stockholder. She retained a phenomenal 62.6 percent equity. Very few entrepreneurs, male or female, can boast that kind of financial track record.

Sandra Kurtzig carved out this generous niche for herself in a particularly male domain. Her company, ASK Computer Systems,

Inc., of Los Altos, California, specializes in software systems for small-to-medium-sized manufacturing companies. ASK's so-called ManMan (short for "*man*ufacturing *man*agement") Information System is a family of integrated programs that run on Hewlett-Packard and DEC minicomputers. They monitor everything that happens in a typical factory—from inventory and work in progress to accounting, purchasing, and production-scheduling details.

ASK sells its software in three ways: either outright, on a time-sharing basis, or as part of a "turnkey system" that includes both hardware and software. To offer its turnkey system, ASK buys minicomputers at discount rates and resells them, accompanied by its programs, ready for immediate installation. Plug it in, load in the programs, and presto! It works.

ASK came about more by accident than design, even though Kurtzig's inquiring mind, frenetic energy, and in-depth technical education (B.A. in math and chemistry from UCLA; M.S. in aeronautical engineering from Stanford) made some kind of career triumph seem almost inevitable. But little more than motherhood seemed inevitable back in 1972 when Sandra quit her computer operations job at General Electric to raise a family and run a part-time contract programming business out of the couple's Mountain View apartment. Two sons followed in short order and they've grown at about the same rate as her company—fast.

In 1974, Sandra Kurtzig incorporated her business with the name ASK—an acronym combining the Kurtzigs' initials, *A*rie and *S*andra *K*urtzig—and started developing programs specifically for manufacturers. It wasn't long before Tymshare Corporation got wind of her expertise and arranged to license her programs for use on its time-sharing network. Because Arie Kurtzig worked for Hewlett-Packard as a research manager, it seemed natural for Sandra to format her software for its 3000 minicomputer series and later to arrange to buy the hardware at wholesale and sell it along with her programs.

To this day, ASK remains debt-free, continuing Sandra's habit of funding expansion with retained earnings. She's equally cost-conscious about her life-style, which hasn't changed drastically since her net worth soared.

In 1982, when an interviewer asked Arie Kurtzig about his wife's material aggrandizement, he shrugged it off. "It's something you

ASK Computer

Sandra Kurtzig's ASK Computer Systems of Los Altos, California, specializes in programs for small-to-medium-sized manufacturing companies. She is one of the few women not only to found a thriving software firm but also to retain a majority equity (in her case, 62.6 percent) after the company went public.

learn to live with," he said and changed the subject. One year later, the couple were divorced. Under California's community property law, the settlement gave Arie $20 million of his ex-wife's fortune.

Sandra doesn't talk about the divorce and is self-deprecating about her success.

"I keep thinking I've reached my level of incompetence but the board won't accept my resignation," she jokes.

The ounce of truth in that statement made her attend Harvard Business School's Advanced Management Program to validate her credentials. She claims it taught her to be less conservative and to take more risks.

In a serious vein, she says, "I'd like to stay CEO of the company, but only as long as I'm doing a good job. Because of my percentage of ownership, I could have recruited a board full of yes-men, but I didn't want that. I wanted board members with diversified backgrounds in marketing, finance, venture capital, and entrepreneurial management—respected people capable of accurately judging my performance.

"I think the major attribute I possess is that I'm able to hire and motivate very good people. There's a lot of employee camaraderie around here, and that's important to keep ASK out in front of the competition."

The Systems Software Savants

"Applications software" is what most laypeople picture when the word "software" is mentioned. For good reason. Applications software—whether for mainframe, mini-, or microcomputers—refers to the programs that actually perform some task for the user. Applications software, for example, makes it possible for people to write letters, store mailing lists, balance their checkbooks, do statistical projections, and even play video games on the computer. The better the applications software, the more user-friendly a computer is said to be.

But that's not the only type of software crucial to the functioning of a computer. Without "systems software," computers would be unable to function.

Systems software operates like a top-notch housekeeper in a palatial mansion. It directs and organizes everything pertaining to the smooth running of the electronic household. It tells the computer how and when to think and where to store its memories. It schedules the computer's many tasks, indicating the order in which it will perform them, and likewise gives the computer's many appendages —its peripheral equipment—their marching orders. In other words, this "operating system" (the type of systems software we are discussing) calls into play all the manifold resources of an electronic household and gets them performing to their fullest. Without such an operating system, a computer would be about as helpful as a houseful of servants with no prior experience, no livery, no incentive to work, and submoronic IQs.

One reason the average person knows little or nothing about systems software is because it is often physically embedded in a computer. The user never touches it as he or she does the disks containing applications software. But since the advent of microcomputers, this is changing. Many of today's desktop models arrive accompanied by a separate disk containing their computer's operating system. Users must load that operating system into their computer at appropriate times just as they load in applications programs.

When the operating system called CP/M was first developed in 1975, it was the only one available for the new microcomputer toys that hobbyists were buying in droves. It is also the first operating system on a disk—in short, the first *portable* operating system the world had ever seen.

With few competing systems in those early years, CP/M (an acronym for "*C*ontrol *P*rogram for *M*icrocomputers") quickly became the de facto industry standard for eight-bit microcomputers. By 1982 there were more than six hundred software companies selling approximately three thousand application packages for use on some 250,000 CP/M-based microcomputers.

CP/M's author, Dr. Gary Kildall, was a thirty-two-year-old professor of computer science at the Naval Postgraduate School in Monterey, California, when he was retained by Intel as a consultant to write a program language for Intel's new and wondrous eight-bit microprocessor. Using PL/M, the language he'd helped develop, Kildall did some further noodling on his own and came up with his initial version of CP/M. He showed it to Intel's management, figuring he could get them to back its development. They declined.

But Kildall and his wife, Dorothy McEwen, also a computer programmer, persisted and eventually offered the system for sale through word-of-mouth and through their own mail-order advertisements. They first realized they were onto something big in 1975 when several of the early microcomputer manufacturers licensed CP/M for use on their machines.

Once the computer hobbyists discovered the system, the Kildalls found that order fulfillment and answering users' questions about

Digital Research

Gary and Dorothy Kildall, husband and wife, pose in front of the original Digital Research building, a restored Victorian house near the sea in Pacific Grove, California. Dr. Gary Kildall is the author of CP/M, the first popular operating system for microcomputers.

CP/M—in the lingo of the computer trade, the latter is referred to as "customer support"—was monopolizing more and more of their time. So in 1976, Kildall quit his job to form their company, Digital Research, Inc.

The couple headquartered the company in a ten-by-thirteen-foot playroom in their small cottage home in Pacific Grove, a picturesque village about 125 miles south of San Francisco. Five years later they were still operating their business, with sales of over $10 million, out of their home in the same seaside town on the Monterey peninsula. But the home was now a rambling yellow Victorian manse with a large carriage house substituting for a production plant. The large house was a monument to their newfound prosperity.

In 1981, the Kildalls accepted an undisclosed amount from four venture-capital firms in exchange for equity in the privately held Digital Research and two seats on the company's board. The Kildalls maintain they didn't need the money, having financed expansion out of profits, like many of their counterparts. But they finally succumbed to the venture capitalists' blandishments when it occurred to them that the financiers' expertise probably was more valuable than their money.

Gary Kildall, more gadget-prone and hobby-prone than business-prone, claims he's not by nature a competitive person. He proved it by the tactical error he made when IBM executives came calling to discuss a deal for making Digital Research's CP/M-86 operating system the standard for the forthcoming IBM personal computer. Where was Kildall when they arrived? Well, it was perfect flying weather that day, you see, so Kildall was off navigating around in the clouds in his private plane, his favorite pastime. Even the weight of IBM couldn't bring him down to earth.

What happened next brought him down to earth with a crash. IBM contracted with a smaller operating systems producer in Bellevue, Washington. On the strength of this one IBM contract, that smaller firm has grown larger than Digital Research. And it threatens to grow much, much larger.

Digital Research's venture backing and concomitant professional management are forcing the Kildalls to run the company in a more traditional manner. To that end, the Kildalls moved the company into an ordinary modern office building. In planning the move, Gary Kildall's only hard-and-fast rule was that Digital Research must, no matter where it relocates, retain a view of the ocean. No doubt that's so he can continue to monitor the flying weather.

Microsoft Corporation is Digital Research's sole independent competitor and the happy recipient of the IBM business that DR flubbed. The only other producers of systems software are a few microcomputer manufacturers. Apple and Tandy Corporation's Radio Shack develop their own proprietary operating systems that they hard-wire into their machines.

Like Digital Research, Microsoft develops and licenses operating systems, programming languages, and other productivity tools for microprogrammers. In addition it just recently branched out into applications software and games programs.

Before the development of MS/DOS, the IBM operating system, Microsoft's best-known and best-selling product was a version of the programming language called BASIC (*B*eginner's *A*ll-purpose *S*ymbolic *I*nstruction *C*ode). It's the product that placed the company on the high-tech map—and eventually attracted a windfall in venture capital. Microsoft's two young founders accepted the financial backing, ostensibly for the identical reason as the Kildalls—the backers' professional expertise in managing rapid-growth companies.

Microsoft was started in 1974 as an informal partnership between two college dropouts in their early twenties.

William H. Gates III and Paul Allen had gone to a Seattle high school together. Then and now, the main thing they have in common is computers. True "droids" in the most obsessive sense of the word,

Microsoft

William H. Gates (seated) and the bespectacled Paul G. Allen dropped out of college to start Microsoft. The company began as a purveyor of operating systems, computer languages, and other productivity tools for microcomputer programmers. Now it also sells applications programs and computer books in its bid to become the country's largest producer of microsoftware.

they've never been able to keep their hands and minds off computers, so it was probably preordained they'd one day make money off them. Even in high school, they'd managed to make some extra pocket money from their mutual passion when they formed something called Traf-O-Data, a DP firm specializing in the computer analysis of traffic patterns. Traf-O-Data got one major job: The Bonneville Power Administration hired it to computerize its electricity grid.

Gates and Allen put in two years each at Harvard and Washington State University, respectively. But the siren song of the microcomputer proved too great for either of them to resist and the pair left college to team up again, this time to write software specifically for MITS's Altair, one of the first commercially produced microcomputers built around the Intel 8080 microprocessor.

In 1975 Microsoft BASIC was ready for shipment, and over the next five years the duo sold over five hundred thousand copies, or more than $4 million worth of it. In the meantime, they continued to enhance the product to make it compatible with an ever wider variety of machines.

As their business grew, doubling in sales every year, they hired other young programmers to develop new products. Often Microsoft's development team worked in conjunction with hardware manufacturers.

Microsoft actually created three operating systems, but one is the money-maker. MS/DOS (*M*icro*S*oft *D*isk *O*perating *S*ystem) has become the de facto industry standard for sixteen-bit microcomputers, just as CP/M is for eight-bit machines. MS/DOS is the system that's licensed for use with every IBM-compatible microcomputer, and there's a growing list of them. Even Apple Computer now offers an IBM-compatible model.

The company also offers microcomputer versions of seven popular languages. Among them are COBOL (*C*ommon *B*usiness-*O*riented *L*anguage), Pascal (named after the seventeenth-century French philosopher and mathematician), and FORTRAN (*For*mula *Tran*slation).

But Gates, who maintains a fanatical commitment to Microsoft working ninety-hour weeks, thinks the future of the company lies elsewhere—in applications software. After all, there will come a time when the market is saturated with microcomputers and most of the new hardware sales are replacement sales.

But applications programs are like phonograph records or cassette tapes. You own one stereo, but you keep on buying records and cassettes to play on it, don't you? The same goes for microcomputers.

PART THREE

Telephones for Tomorrow

5 Conveying a Zillion Conversations: *Ma Bell and Competitors*

Of the many ironies surrounding the history of the American Telephone and Telegraph Company (AT&T), perhaps the most pointed involves the man with his picture engraved on the company's stock certificates. That picture of Alexander Graham Bell, the inventor and owner of the telephone patent that started it all, may be as comforting to investors as that of George Washington on our dollar bills. It's unlikely, though, that the likeness has the same effect on Bell's heirs. True, the descendants of Alexander Graham Bell aren't starving; one branch still owns *National Geographic* magazine. But neither are they enshrined in the pantheon of the superrich.

Surprised? Don't be. Throughout American history, invention has never provided a particularly well-traveled pathway to riches—often because the inventor, like Bell, is more interested in frittering away the proceeds on more tinkering than in salting it away for posterity.

How rich was Bell? In comparison to the likes of the two seniors—John D. Rockefeller, oilman and Henry Ford, car mogul—Alexander Graham Bell's genius, certainly equal to theirs, did little more than leave him comfortably well off—upper middle class, you might say. The reason was that Bell had almost immediately sold off the bulk of the family's Bell Telephone Company stock, retaining only two thousand shares. He cashed in soon after the company's founding in 1877 during the initial bull market for the shares.

Alexander Graham Bell had started out with a 30 percent interest in the company that became the predecessor firm to AT&T. It is no doubt cold comfort to his heirs that patent No. 174,465 is frequently termed the single most valuable patent in the nation's history.

For almost half a century, AT&T reigned as the world's largest corporation. But after January 1, 1984, AT&T reigned no longer. That is the date this dinosaur monopoly became extinct. But it was reborn in the guise of eight separate legal entities. Seven of those entities continue to be regulated by state and federal authorities. The eighth is unregulated except for one of its divisions, which will receive directives from the Federal Communications Commission (FCC).

Where once there were twenty-two "operating companies"—the companies bearing names such as Mountain Bell and New England Telephone that we all came to love or hate as the sole local provider of our telephone service—after the 1984 divestiture, there were seven regional holding companies. The eighth surviving company is AT&T itself, which retains the leftovers from its former empire.

But what leftovers they are! AT&T has reorganized into two gigantic divisions: The regulated division, called AT&T Communications, absorbed the old Long Lines subsidiary and continues to provide long-distance telephone service. The unregulated division is called AT&T Technologies. It handles telephone and computer equipment research, manufacturing, and sales. It encompasses Bell Laboratories, perhaps the country's premier high-technology research organization; Western Electric Company, the former equipment manufacturing facility; AT&T International, selling its telecommunications expertise abroad; and a newly formed subsidiary christened AT&T Information Systems.

AT&T Information Systems represents the real departure from telephone company tradition. Thus it absorbed most of the kicking and screaming that accompanied the whole divestiture idea. The controversy concerned AT&T Information Systems' *raison d'être:*

The subsidiary's stated objective, since AT&T officers first broached the subject years earlier, is to become a broad-based "information company" offering many of the same computerized products and services as the biggest of them all, IBM.

The notion of AT&T as a "knowledge company"—what its ads now proclaim it to be—started to dawn on company officials in the 1970s as computers and breakthroughs in fiber optics technology increasingly blurred the distinctions among voice, data, and image transmission. AT&T masterminds reasoned that the company was in the computer business in any case, since it leased many of its lines for the transmission of electronic data. Unfortunately, AT&T couldn't act like it was in that business—or in many of the other pioneering businesses it wanted to be in—because it was a regulated monopoly.

And that wasn't all AT&T officials had to say. What about Bell Labs? That innovative R&D think tank had launched the whole microelectronics revolution with its invention of the transistor and other state-of-the-art devices. But there were only two ways AT&T could capitalize on the imagination of its scientists: It could use their breakthrough ideas to improve its telephone equipment, or it could license its inventions to outside corporations, usually for less than their true market values, as mandated by a 1956 consent decree. Naturally, when it did the latter, its licensees made the colossal bucks, profiting mightily from AT&T scientists' forward thinking.

Texas Instruments, for example, obtained a license for AT&T's transistor technology in the 1950s for under $250,000. For this trivial sum, Texas Instruments was able to transform itself from an oilfield instrument firm called Geophysical Services, Inc., into a multibillion-dollar semiconductor company, No. 1 in that industry.

In short, AT&T wanted the same maneuvering room as its competitors. It wanted to be able to compete in the information marketplace of the future on an equal footing with all comers. The breakup plan was the compromise AT&T officials finally reached with the Antitrust Division of the United States Justice Department and the FCC in order to breathe life into their dream.

AT&T's dismemberment is unleashing competition in areas ranging from long-distance telephone service to equipment manufacturing. But the real riptide had already erupted a decade or more ago because of a determined Texan named Thomas F. Carter.

Today the semiretired Carter is the most famous resident of Gun Barrel City, Texas, where he moved in 1975, even though he's long since faded from the national limelight. However, the name Thomas F. Carter is one Ma Bell will never forget—nor will the authors of American business history.

In the 1950s, when Tom Carter was running a two-way mobile

radio business in Dallas, he invented a clever little device that allowed his customers to hook directly into the local telephone system. Customers could sit in their cars and actually talk to people on the other end of a regular telephone line. To accomplish this minor feat, the customer radioed to his base station which, in turn, plugged him into the phone system using the clever little device, termed the "Carterfone."

The catch was the local Bell subsidiary, which got wind of Carter's operation and forbade him to attach the Carterfone—or any non-Bell equipment, for that matter—to its lines. Carter filed an antitrust suit in 1965 against both AT&T and the United States' second largest independent phone company, General Telephone and Electronics Corporation (GTE). The suit was referred to the FCC, which, in 1968, ruled in Carter's favor, to everyone's utter amazement.

Almost immediately, "the Carterfone decision" gave birth to a bustling industry of combative Bell System competitors. And it was this newfound "interconnect industry" of pugnacious telecommunicators, vociferously accusing AT&T of trying to stifle competition and put them out of business, that eventually led to the federal government's 1974 antitrust suit and the leviathan divestiture of 1984.

These upstart companies reaped the benefit of Carter's persistence, but Carter himself never seemed able to exploit his victory—at least not on any grand scale. It is rumored that Carter received more than $300,000 to settle his original antitrust suit. But that and the satisfaction of appending his name to a landmark decision were about all he did get.

Even after the shackles were removed, his Carterfone never took off. The reason had more to do with basic telephone etiquette than anything else.

Back then, mobile radios were designed to be shared by multiple users. It was somehow understood that mobile radio users should limit themselves to short, specific messages. But the psychology of the telephone call is different. Telephone conversations, by their very nature, tend to ramble as the two parties exchange pleasantries before getting on with the practical purpose for the call if, indeed, there is one. So when people used Carter's mobile radio to chat on the telephone, it tied up channels for extended periods and made other users hopping mad.

By 1970, when some of AT&T's latter-day rivals were just getting under way, Tom Carter—the lone combatant who had made it possible for them to go into business in the first place—had stopped making his Carterfone and left his company. His Carterfone Communications Corporation is now a division of Cable and Wireless Ltd., a British communications giant. The division sells and repairs teletypewriters and computer terminals.

But Carter, the old warrior, wasn't through making trouble for Ma Bell. After starting a few more businesses that didn't last long, Carter founded the North American Telephone Association, an industry trade group representing non-Bell telephone equipment makers. His lobbying organization would eventually encompass some 350 companies ranging from old-line ITT to the feisty upstart ROLM Corporation.

Among Tom Carter's later ventures was a company called Florist Transworld Delivery system, which sent flowers by wire. His partner was an enterprising midwesterner named John D. Goeken.

A decade earlier, in 1963, Jack Goeken had tried to launch a regional mobile radio company similar to Carter's called Microwave Communications, Inc. (MCI). He, too, had slammed up against a Bell System stone wall when he tried to get the FCC's permission to build a small microwave system to service truckers along the main highway between Chicago and St. Louis. Bell's opposition was vigorous even though Goeken claimed his CB-radio-based system was not intended to compete with AT&T.

AT&T was not the only company filing an objection. Arrayed against Jack Goeken and his four silent partners—all friends from Joliet, Illinois, where MCI was based—was a communications juggernaut comprised of Illinois Bell, Southwestern Bell, GTE, and Western Union.

"Their whole hope was, boy, this guy is going to run out of money," Goeken recalls. "They came in and said, number one, that the system isn't going to work—God liked them and only let microwave work for the Bell System. Number two, there was no market for it. And three, that nobody would give us the financing to build the system. Well, now, if they really honestly believed all that, then why would they spend a nickel fighting us?"

With such monumental opposition, Jack Goeken's idea languished in the file cabinets of the FCC until 1968, when a tenacious entrepreneur-turned-venture-capitalist named William G. McGowan entered the picture. Goeken couldn't have found a better person to challenge the hegemony of AT&T, for Bill McGowan's tenacity and disregard for the established order of things are due as much to heredity as environment.

Hailing from the same coal-mining region of Pennsylvania where the Irish immigrant labor organizers known as the Molly Maguires had made life difficult for their bosses in the nineteenth century, Bill McGowan's father had continued the tradition in the twentieth century as a railroad union organizer. The same anti-establishment spirit was passed on to McGowan the younger, who grafted onto his progressive point of view an excellent technical education (B.S. in chemical engineering) that culminated with an M.B.A. degree from Harvard Business School. At Harvard he was selected a Baker Scholar, meaning he was in the top 5 percent of his class. He graduated in 1954.

McGowan, who looks a little like the old-time movie comedian Joe E. Brown, began his career with the Magna Theatre Corporation where he worked with Mike Todd, observed his bravura business style, and helped develop the revolutionary Todd-AO process for projecting films onto a wide screen.

Four years later McGowan was ready to flap his entrepreneurial wings, and he launched the Ultrasonic Corporation in Hicksville, Long Island. With $50,000 capitalization, his company developed cryogenic measurement devices for the United States space and missile programs. Within a year, based on its agreements with several large aerospace contractors, Ultrasonic became a government subcontractor.

While McGowan's venture prospered, it did hit an occasional snag. For example, in the early 1960s, the company built a prototype for an ultrasonic device intended to repel sharks. McGowan figured it was a perfect contraption to unload on the U.S. Navy and invited the head honchos to witness a demonstration. Unfortunately, the sharks were more curious than afraid of the device and started gnawing away at it. Displaying a characteristic quick wit and showmanship picked up from Todd, McGowan flashed his broad Joe E. Brown grin and announced he'd actually invented an aphrodisiac for sharks that would distract them from eating people.

Bill McGowan's unfailing sense of humor and ability to see the bright side of every setback would stand him in good stead in the telecommunications battles to come.

McGowan made his first million when Ultrasonic went public; it was acquired in 1963. For the next five years of his life, he set up shop as a consultant/venture capitalist reviewing other people's ideas about how to get rich. McGowan remembers it as one of the most pleasant and hassle-free times of his life:

"I was really enjoying myself," he recalls. "I was intrigued with new products and services and attracted to ideas which hadn't been put into effect before. Frankly, I enjoyed the challenge of doing things that didn't have a set pattern. As people became aware of my interests, they came to me with different opportunities. I helped these businesses with their funding and organizational structure."

Jack Goeken's stalled FCC application was one of the projects that caught McGowan's attention in 1968. Microwave Communications, Inc. (MCI) contained all the elements that appealed to McGowan. Despite the Carterfone decision, the project seemed impossible on the face of it. That element alone made McGowan, a born gambler, drool. After all, what could be more of a long shot than whipping Ma Bell?

Soon Jack Goeken and his partners and Bill McGowan had struck a deal : McGowan would take over MCI's management and find a way to pay off its debts in exchange for 25 percent equity.

But once McGowan was in the driver's seat, it became apparent that the old boss Goeken and the new boss McGowan were fixed on a collision course. Unlike Goeken, whose interest is devising pioneer uses for existing telecommunications technology, McGowan saw MCI as a national telephone company serving the business community.

Although Jack Goeken let McGowan proceed with his crusade against Ma Bell, the difference in the two men's goals would eventually cause Goeken to leave the company in 1973. But Jack Goeken did not leave a beaten man. He stormed out with 1.5 million MCI stock certificates tucked away in his safety deposit box, equity that, over time, grew into a personal fortune rivaling McGowan's.

Outlining MCI's early tribulations, McGowan once quipped: "First, we were a venture-capital operation, then a lobbying company, and

MCI

In 1968, William G. McGowan was a successful entrepreneur turned venture capitalist when he got involved with a foundering telecommunications firm called Microwave Communications, Inc. McGowan helped transform MCI into the country's second-largest provider of long-distance telephone service after AT&T.

for five years we also had to act as a law firm."

MCI had to act as a law firm to win its much-heralded 1978 victory over the behemoth Bell in the federal courts. "Legal R&D" is McGowan's term for the whole lengthy, jurisprudential process the company would eventually endure.

Actually, McGowan's initial money-raising technique for his national telephone enterprise was as inspired as his subsequent fancy legal footwork.

First, Bill McGowan formed seventeen regional operating companies. Each of these new companies—sporting names such as MCI New York West and MCI Texas Pacific—applied to the FCC separately for permission to build a leg of the MCI microwave network-to-be. Each company also had its own pool of investors who, *in toto,* kicked in $7 million in collective start-up funds. MCI retained 30 to 40 percent equity in each operating company and acted as a consulting firm. MCI collected fees for services rendered to each of these seventeen affiliates. These services ranged from developing customer leads to preparing and following up on all FCC filings.

In one of his first official acts, McGowan had moved MCI's base of operations to Washington, D.C., just so it could act as a kind of trade association-lobbying arm for its affiliates.

In 1969 one of the MCI family of companies won its first FCC approval—for the Chicago-to-St. Louis route for which Goeken had originally applied. Two years later, in 1971, the FCC granted licenses to the rest of the MCI companies, with the explanation that all qualified applicants should be afforded "the freedom to fail" in the long-distance, intercity telecommunications market. Some vote of confidence.

Known as "the MCI decision," the 1971 FCC decision was precedent-setting. Why? Because by rendering its favorable decision, the FCC also approved applications from four other companies to construct domestic satellite systems that would also compete against AT&T's Long Lines Division. The FCC dubbed these Bell System competitors "specialized common carriers."

With the FCC go-ahead, McGowan moved fast. He merged the seventeen affiliates into the parent company, took it public, negotiated credit with five banks, and managed to raise an incredible $120 million, which he immediately plowed into the construction of his nationwide microwave network.

Microwave transmission requires relay towers spaced a maximum of forty miles apart. To blanket the country, MCI obviously needed to build a prodigious number of relay towers. McGowan did it the cheapest way possible. MCI's towers were prefabricated and tested in a plant in Richardson, Texas, shipped to the sites, and erected in a matter of hours.

During MCI's peak construction period in 1973, McGowan claims his workers were erecting them at the rate of one per day. This is a job that usually takes AT&T months, granting that its towers are technological works of art compared to MCI's. The important thing, though, was that MCI's towers worked.

But MCI's service to its customers was spotty and undependable due to Ma Bell's continuing obstinacy. It was fine for MCI to offer its customers dedicated private lines—what we think of as "leased lines" in Bell System parlance—but AT&T still had to cooperate to make the service a reality. MCI could transmit voice signals across the miles via its own microwave relay network with no trouble at all, but at some point those signals still had to enter a Bell System line for the final leg of the journey to a telephone in someone's office.

Actually, AT&T had cooperated on the first Chicago-to-St. Louis route only to watch MCI, with its inexpensive rates, capture over 80 percent of Bell's private-line service between those two cities within twenty months. Ma Bell was horrified, shrieked that MCI was "skimming off the cream" of the long-distance business, and refused to interconnect MCI to any more of its lines. Henceforth, AT&T simply ignored the 1971 FCC directive.

By September 1973, the AT&T roadblock had precipitated a crisis. By this time, MCI had spent $80 million but still had no regular income it could count on to pay off its debts and the fixed expenses on its sprawling network.

Bill McGowan recalls being in a state approaching panic. He got the FCC to write AT&T a stern letter. Nothing happened. He went to the Justice Department, which started an investigation. Nothing happened. Then MCI filed its own antitrust suit. All well and good, except that this method of redress, even if it worked, would take years. What MCI needed was immediate action.

On the brink of extinction, MCI finally got its reprieve on December 29, 1973, when a federal court in Philadelphia issued a writ directing AT&T to order its employees to interconnect MCI with its customers. The cliff-hanger was over. Grudgingly, Ma Bell and her children complied, although McGowan maintained during testimony at MCI's antitrust suit trial in 1978 that Bell was never all *that* cooperative.

MCI's 1978 victory in that later antitrust case created far more than just a ripple of reaction. It sent waves crashing through the telecommunications-industry ocean.

As most interpreted the court's decision, from then on any entrepreneur wily enough to gain FCC approval could enter the competition against AT&T's Long Lines Division, offering the entire range of long-distance telephone services.

For MCI, the 1978 decision cut two ways. For its trailblazing, MCI was rewarded with its own pack of imitators nipping at its heels. These "WATS line resellers," as they are dubbed, flew into the breach in the early 1980s. True, many died a natural death after a year of trying, but others, such as U.S. Telephone Communications in Dallas and Combined Network in Chicago, threaten to bite off a sizable chunk of MCI's business.

In the 1980s, MCI's many hard-won triumphs over AT&T are paying off to the point where McGowan jokes that MCI is really an acronym for "*m*oney *c*oming *i*n." In 1977 MCI turned its first profit, and in 1980 it began wooing residential customers for its long-distance service with success.

But despite rates 20 to 50 percent cheaper than Ma Bell, MCI and the "reseller" crowd do have their detractors. They complain of inferior sound quality and bills riddled with mistakes. (One wall AT&T has managed to defend is the one securing its superior billing technology, which it steadfastly refuses to sell to the likes of MCI.) In addition, during periods of heavy demand, MCI customers can expect up to five times as many busy signals as they'd get using Ma Bell's Cadillac service.

Apparently plenty of people feel they can afford the Cadillac, because by 1982 AT&T's Long Lines still had 96 percent of the domestic long-distance market. But MCI had captured a respectable 2 percent of it. And in 1982 this market share helped MCI push its revenues across the $1 billion mark for the first time.

Never content to rest on his laurels, Bill McGowan is plowing MCI's profits back into the business and adopting more advanced modes of telecommunications delivery. Even though MCI has no research or elaborate production facilities of its own, McGowan has always stayed informed about sophisticated telecommunications technology and encouraged manufacturers of state-of-the-art devices to demonstrate them to him.

Thus it's hardly surprising that MCI is one of the first American telephone companies to install fiber optic cables. Similarly, MCI purchased twenty-four transponders, or channels, on two Hughes Communications satellites launched in 1983 and 1984. And it bought WUI, Inc., from Xerox Corporation to give it instant access to the international telex, data-communications, and cable-message market and the exploding domestic mobile-radio and pocket-paging business.

Finally, it is leading the thrust to bypass the deteriorating and out-of-date facilities of the Bell companies altogether and instead use cable television wires to move messages between residential customers and its network.

Phones That Roam Through a Cellular Honeycomb

Radiotelephone service to and from vehicles first drove onto the scene in 1946. It is a technology developed by and licensed to the Bell system companies.

Since its inception, there have always been far more people who wanted the service than there were FCC-allotted radio channels to accommodate them. And this was despite the fact that this service costs up to twenty times as much as residential service.

As of 1980, for example, there was a waiting list of more than 40,000 would-be users. This is over and above the 130,000 vehicles in the United States that were outfitted already with radiotelephones. Industry observers say this waiting-list figure is an impressive indicator of pent-up demand for several reasons:

Because many people don't bother to get in line if they know there's a long wait, these observers speculate that demand is really in the hundreds of thousands. Moreover, they project it will be in the millions as soon as a new and better radiotelephone technology hits its stride in the late 1980s.

The new computer-based technology is called "cellular radiotelephony." It is a vast improvement over the old car telephone service in terms of both its quantity and its quality.

Using the older, conventional mobile telephone technology, not only is the sound uneven, but also the wait before users can get an open radio channel on which to make their call is usually very long.

Take New York City as an example. Before the new cellular system was introduced there in 1983, there were only twelve channels available for New York Bell's 700 lucky mobile radiotelephone customers. But were they lucky? Since only twelve conversations could take place simultaneously, the other 688 customers had to drum their fingers on their dashboards in the meantime—or, more likely, watch their chauffeur drum his fingers, since most of those 700 phones were in limousines.

With the new technology, all that is changed. Using the cellular system, there's only a 2 percent chance that a customer can't make his call on the first try—despite the fact that the new technology will enable New York Telephone to make the mobile telephone service available to a maximum of 250,000 metropolitan area customers. And in addition to eliminating the wait, the quality of cellular voice transmission rivals that of regular telephone service.

How is this revolution in mobile phone service possible?

Instead of one high-powered radio transmitter with a twenty-to-twenty-five-mile range, a cellular radiotelephone system, or network, blankets a locality with many shorter-range transmitters. Each of these smaller transmitters sits in the middle of a "cell," which is a geographic area about one to eight square miles in size. The cells form a honeycomb pattern.

Every radiotelephone network has a limited number of channels, or frequencies. These channels are divided up equally among the cells, the main requirement being that no two adjacent cells use the same channels. Thus the same channels can be reused many times by other nonadjacent cells in the network's honeycomb pattern without any interference problem.

Each cell's individual transmitter sends and receives signals from customers' small computerized car telephones. In turn, all the cell sites are wired to a large computer in a central switching office that tracks the car phones as they roam through the streets and highways.

As a customer in a moving vehicle talks, the central computer is continually "locating" available channels in adjacent cells and assessing when to "hand off" that call to a different channel in the next cell. Through the miracle of modern computer technology, this transition—or hand-off—is accomplished so swiftly and silently that the people speaking to each other never know it happened.

Cellular systems accommodate growing customer loads through a kind of electronic mitosis. Cells are simply subdivided into smaller and smaller units with incrementally lower-powered transmitters.

The cellular radiotelephone proponents say that car phones are just the beginning. The same cellular technology can as readily support pocket-sized portable telephones and even tiny wristwatch versions someday.

Cellular radiotelephony was developed by AT&T over a thirty-year period at a cost of more than $200 million. The Bell operating companies actually applied for the first licenses way back in 1968. Since then, the FCC has been sitting on the technology.

To alleviate some of the bottled-up demand for car telephones, though, the FCC in the 1970s did allote more frequencies for that purpose. Some went to Bell and the independent telephone companies (known in FCC jargon by the descriptive term "wire-line common carriers" or WCCs).* Others were assigned to the country's approximately 750 specialized, privately owned mobile telephone companies (known since 1949 as "radio common carriers" or RCCs).

But it wasn't until May 1981 that the FCC finally decided about the dissemination of the new cellular radiotelephone technology. It was a decision redolent of the wisdom of Solomon:

To break the logjam of demand, the FCC would allocate forty frequencies for cellular service in each radiotelephone market. Twenty of those channels would be granted to a wire-line CC—in other words, a local telephone company—and the other twenty to a designated RCC. The FCC would accept applications to operate cellular systems in the nation's top thirty markets until the deadline of June 1982, after which it would deliberate and probably begin issuing the first licenses within the year. Applications for the remaining smaller markets would follow and be staggered.

*Besides the Bell companies and GTE, there are approximately 1,760 small independent telephone companies in the United States.

The FCC made two final points: It would be the most favorably disposed toward those applicants who formed joint ventures with competitors and applied collectively for one license. And no additional frequencies in the radio spectrum, which is finite, were being reserved for any future growth of cellular systems.

For would-be cellular radiotelephone entrepreneurs, the latter pronouncement was the most significant. The FCC's message to them came through loud and clear: Either get your cellular license now or be shut out of the market forever.

One FCC commissioner characterized the stampede for these first sixty franchises as an example of capitalism at its most excessive, a California Gold Rush mentality.

"I've been in government for twenty-five years, and I didn't think there was a scintilla of innocence left in me," said James Fogarty. "But I've been amazed by the greedy reaction throughout the communications industry to the prospect of getting cellular licenses."

When the dust settled after the June 7 filing deadline, the seven FCC commissioners had to decide which of the almost two hundred applicants would be given the opportunity to become rich.

Barry Yampol, for one, wasn't taking any chances. As the founder and head of Graphic Scanning, one of the largest radio paging companies in the country and headquartered in northern New Jersey, he was determined to reap his share of licenses. It took two semitrailers to deliver Graphic Scanning's voluminous proposals constituting its more than seventy-five applications. In those proposals, Yampol claimed he could deliver the highest-quality service over particularly wide areas at a cut-rate cost about $20 to $30 per month lower than the charges projected by most of the other applicants.

Naturally, Yampol's rivals were incensed. It's a reaction Yampol is inured to as an aggressive young entrepreneur who went from near zero personal worth in 1968 when he started his company to over $80 million in 1982.

Graphic Scanning is a publicly held company with interests in a number of fast-growing communications services such as electronic funds transfer, telex, and electronic mail. Yampol files the requisite 10-K forms with the Securities and Exchange Commission but refuses to publish annual reports or otherwise communicate with the investing public. He feels his 24 percent equity gives him the right to be secretive.

"My business is not selling stock," he once snapped to a reporter. "My business is selling services."

Unfortunately for Yampol, the aspiring cellular Croesus, doubt was cast on his probity, not to mention his credibility, a couple of months after the FCC deadline. His competitors accused Graphic Scanning of setting up four shell companies to apply for hundreds of paging licenses without disclosing its interest in them. Yampol denied everything and welcomed the FCC investigation. Yampol claims his engineers often undertake surveys that other RCCs incorporate in their FCC applications for paging licenses. Why not? It brings extra consulting fees into Graphic Scanning's coffers.

While the RCCs were fighting among each other for the cellular franchises up for grabs, the WCCs were quietly and efficiently forming joint ventures just as the FCC had requested. As a consequence, the cellular licenses earmarked for telephone companies were awarded almost immediately, giving AT&T, GTE, and the eighteen small local companies in the WCC consortium a head start in the marketplace.

In October 1983, the first Chicagoan—indeed, the first person anywhere in the United States—made a phone call from his car via a cellular mobile radiotelephone. The man made a long-distance call to New York. The company offering the service was a limited partnership comprised of Ameritech, a Bell regional holding company, and Centel Corporation, an independent telephone company operating in the Chicago area. Ameritech owns 93 percent equity, Centel owns 7 percent.

Phones with Feet That Go Beep-Beep

One-way beeper paging units have been used for years to alert physicians and equipment repairmen on call that they've got an urgent message waiting. The beep-beep tells them to head for the nearest telephone to call a special phone number to retrieve their message.

Until recently, pagers that go beep-beep were no respecters of other people's rights. Pagers also alerted everyone else within earshot, turning many a doctor's face beet red with embarrassment in the middle of a live theatrical performance or during that tense moment on the putting green just before a friend's club makes contact with the ball.

Advances in paging technology have finally eliminated such awkward situations. Instead of a disruptive beep-beep, pagers now have other methods of getting your attention. Some vibrate. Others vibrate and then print out a short message or phone number on a small display screen similar to that of a pocket calculator. Even more startling are the pagers that beep or vibrate to get your attention and follow up with a brief voice message.

But the most miraculous of all are the pagers now in the prototype stage that physically print out a message on a small roll of paper tape. Such capabilities will transform pagers into pocket data terminals, able to supply doctors with instant medical histories, repairmen with product serial numbers and maintenance records, investors with up-to-the-minute stock quotes, and consumers with all the personalized news and weather information their hearts might desire and their brains can absorb.

Perhaps the best feature of all is that contemporary pagers, no matter what their capabilities, are lightweight and compact. Most units are no bigger than a cigarette pack. Motorola makes a beeper that's about the size of a fountain pen.

Clearly, these new permutations in paging technology will make the service appealing to far more people, especially America's legion of peripatetic business executives.

Apparently the FCC agrees. In 1982 it voted to make 120 more radio frequencies available for use by one-way paging systems. With as many as 100,000 customers serviced by a single channel, that's room for 12 million customers in addition to the 1.5 million people already carrying beepers.

As usual, the companies already offering paging service are the ones who stand to benefit most from the expected boom. Having already proved their staying power and dexterity in the paging marketplace, they will undoubtedly be the firms the FCC favors with the majority of the approvals to operate on the new frequencies. Several once-and-future Midases will be made in the process, since many of these RCC applicant firms are not that far removed from their entrepreneurial origins:

Although Graphic Scanning long ago diversified beyond paging, Barry Yampol has never been one to let any financial opportunity slip by unseized. He's applied for a number of licenses.

Bill McGowan, too, has put in his bid through MCI's acquisition of Airsignal, a company operating paging services in forty-one mostly medium-to-small markets. On the heels of that acquisition, MCI joined forces with American Express Company, Metromedia, Inc. (the nation's fourth largest broadcaster), and the Dallas-based Communications Industries (one of the top five RCCs) to establish a formidable national telephone paging system that will beam its signals off satellites.

David Post of PageAmerican Group, Inc., a New York paging outfit he founded in 1977, also has stars in his eyes. This ex-Wall Street analyst formed a joint venture with RCA Global Communications to offer a one-way message service that beams its long-distance signals off satellites. The service provides subscribers with information ranging from news headlines and stock quotes to the names and phone numbers of people trying to contact them. Subscribers carry around a pocket-size terminal, dubbed a "PageGram," that displays messages on a tiny alphanumeric screen.

"The PageGram provides the same information that normally reaches a person's desk via a telex machine or computer," David Post explained. "It's the nearest thing to a pocket electronic mailbox. We think our network—and it is a network rather than a product—is going to change the way people live and work. No one will ever have to be out of touch again no matter where they are traveling throughout the world."

Phones That Fly Through an Azure Sky

Jack Goeken, the man who founded MCI, is now pursuing his second pot of gold. His latest enterprise is AirFone, Inc., the only company currently licensed—albeit on an experimental basis—to offer radiotelephone service from airplanes in midflight. The company is a 50/50 joint venture of Goeken Communications, Inc., and Western Union.

Washington, D.C.-based AirFone appears to have everything going for it: It's well financed. It has no competition. And, as of this writing, it has signed up ten of the country's major airlines.

The airlines welcome the service since they receive a 10 to 15 percent commission on the revenues. And the revenues generated could be substantial. The price certainly is. To call anywhere in the continental United States, it costs $7.50 for the first three minutes and $1.25 per minute thereafter up to forty-five minutes, when the signal begins to fade.

Forty-five minutes is the maximum calling time because the new single side-band technology on which the system is based does not have the capability of switching or "handing off" calls from one ground receiving station to the next. It would be prohibitively expensive, Goeken claims, and only the big talkers would benefit.

As the AirFone system exists now, each of its thirty-seven terrestrial stations that dot the country's midsection from New York to California has a receiving range of about two hundred miles. On board each plane outfitted for the service is a small radiotelephone computer. To allow the caller maximum talking time, the computer automatically beams the signal to the station within the two-hundred-mile range that is the farthest in front of the speeding plane. Goeken plans on adding many more earth stations to the network eventually. Only then will AirFone become a truly nationwide service.

6 Switching On: *The Telequipment Tycoons*

Bill McGowan isn't the only entrepreneur who has profited handsomely from AT&T's heavy-handedness over the past decade. There's also Fred Cohen of TeleSciences, Inc. True, it would be hard to find a more reluctant adversary. After all, AT&T has always been one of TeleSciences' best customers.

Cohen and two partners founded TeleSciences as a moonlighting operation in a suburban Philadelphia basement in 1966. At the time Cohen was employed as an engineer at RCA, but he was determined to find an entrepreneurial escape route out of the corporate jungle. TeleSciences was it.

Cohen, a lawyer named Michael Petrozzo, and a young ITT technician, Dony Crowder, each put up $500 and embarked on an ambitious goal—to design and manufacture advanced, computerized telecommunications systems to sell to telephone companies. In the United States in the 1960s obviously that meant selling to AT&T.

TeleSciences' first product was rather simple and straightforward: a switching matrix test set. Western Union, its initial customer, bought it to detect any faulty wiring in its switching office that linked teletypewriters.

Two years after its primitive beginnings, TeleSciences was paid several visits by representatives of the First National City Bank (now called Citibank). The bank representatives smelled a good thing and wanted to invest some pension-fund money in it. TeleSciences finally agreed to accept the bank's check for $1 million in exchange for 36½ percent equity.

Shortly thereafter, the company went public. Cohen and partners used the money to develop TeleSciences' best-selling product AUTRAX.

AUTRAX is a complex piece of equipment that is placed in a telephone company's central switching office to monitor and statistically analyze traffic on the network. It replaces an army of human troubleshooters, for it instantaneously pinpoints malfunctions that previously took many man-hours to locate.

When this traffic-engineering tool, bearing a hefty price tag of over $1 million, was introduced in 1970, the Bell System immediately bought several, since Western Electric couldn't offer anything approaching it. But it wasn't long until Western Electric did start making a similar system and, needless to say, AT&T's AUTRAX purchases slackened. Still, AT&T remained TeleSciences' single largest customer for AUTRAX and its other equipment, accounting for about 25 percent of the company's business. Independent phone companies accounted for another 25 percent, and the remainder was foreign.

Because AT&T was still TeleSciences' largest customer, Cohen was understandably reluctant to sue. He didn't sue until the U.S. Justice Department forced his hand.

During its massive federal antitrust suit against AT&T, Justice Department attorneys subpoenaed TeleSciences' records and cited the company's experience as one of sixteen examples of Bell's favoritism toward its own manufacturing arm, Western Electric. Cohen debated the issue with his advisers—his original partners had long since departed. They eventually decided to file TeleSciences' own antitrust suit in 1980.

TeleSciences' pluck paid off in a way Cohen never anticipated. A year later, to settle the suit, AT&T agreed to purchase $300 million worth of TeleSciences' equipment over a period of eight years. To show good faith, it made an advance payment of $40 million. The amount was a godsend to a company whose total annual revenues were only just approaching the $35 million mark.

The windfall gave Cohen, the owner of 5 percent equity, the confidence to opine that TeleSciences has a shot at making the *Fortune* 500 list someday. But Cohen knows the company will never make it unless it diversifies its product line and increases its sales abroad.

* * *

One of the new products TeleSciences plans to offer is a computerized device sold to corporations that have over fifty in-house telephones. This multiline "private branch exchange," known by the acronym PBX, is the modern version of an old-fashioned company switchboard. But PBXs are fully electronic. A computer does all the work.

There is, however, an older electromechanical switching device still used in today's smaller offices. It's called a "key system." You know your company has a key system if your individual office phone has a row of buttons ("keys") that light up to indicate which lines are busy. To reach the outside world, you just press an unlighted button and dial the number. In contrast, to make a call outside the company on a phone linked to a PBX, you first dial "9." PBX phones may or may not have buttons.

Because key systems are electromechanical, it could take a repairman hours to change the wiring and make the necessary adjustments every time employees move to new quarters or change their extension. Not so with computerized PBXs. Moves, or expansions and upgrades to the system, are accommodated easily by revising the software—in other words, by entering new configuration information into the computer. Monkeying with machinery is no longer necessary.

Many telecommunications analysts question TeleSciences' wisdom in deciding to manufacture PBXs, since the field is already overcrowded, with about forty would-be giant-killers. AT&T still predominates, with about 35 percent of the market, although its postdivestiture market share is a far cry from its overwhelming 90 percent share fifteen years ago.

Second, with about 15 percent, is a young Silicon Valley company named ROLM Corporation.

ROLM is the shooting star of the interconnect industry. It was founded in 1969 by four Rice University alumni—Walter Loewenstern, Jr. (Rice '59), M. Kenneth Oshman and Gene Richeson (Rice '62), and Robert R. Maxfield (Rice '63). All four had been electrical engineering majors, and all four migrated to northern California to work and to get graduate degrees at Stanford. Altthough they hadn't all known each other in Houston, those who had renewed acquaintances in California, and the four of them became good friends.

Not surprisingly, the quartet in time became infected with the entrepreneurial fever that has turned San Jose and environs over the past decade into one vast maternity ward for embryonic companies. The foursome decided it was their turn. They were all about thirty years old. The problem involved what to produce. They began brainstorming.

ROLM

ROLM Corp., the shooting star of the telephone equipment industry, was founded by four Rice University alumni who had migrated to northern California in the 1960s to work and to earn graduate degrees at Stanford. Gathered for a rare group portrait are (left to right): Gene Richeson (no longer with the company), Bob Maxfield, Walter Loewenstern (in the white suit), and Ken Oshman.

Most of their ideas were commercial applications for computerized systems used by the military and other government authorities. For example, how about a system that would make it possible for police headquarters to track twenty-four hours a day the whereabouts of all their squad cars? Or how about a system that enables toll bridge authorities to monitor the regular users of the bridge automatically by means of a transponder attached to the person's car, thus permitting monthly commuter billing?

With such pie-in-the-sky ideas, venture capitalists did not flock 'round.

Finally, Gene Richeson suggested that what the world really needed was a low-cost, off-the-shelf military minicomputer. He was working for a company doing advanced military systems engineering so he understood something about the market and the technological requirements.

A "mil-spec" computer has to be rugged to withstand adverse environmental conditions—temperature extremes, explosions, sand, salt spray, and electronic interference. For this reason, mil-spec computers are always more expensive than comparable non-"ruggedized" computers. In addition, up to that time military computers were all custom-designed mainframes with long lead times.

But when the foursome got interested, the year was 1968 and minicomputers were just coming into their own. At a computer trade show, Richeson and Maxfield had been favorably impressed by Data General's mini, dubbed NOVA. Since Data General was a new company, it might welcome offers to license its technology, they speculated. So why not license the design for NOVA and reengineer it to military specifications?

The beauty of their NOVA idea was that their mil-spec machine would be compatible with Data General's cheaper commercial model. Thus the user could develop and debug the machine's software on the commercial model in a benign laboratory environment and use only the special mil-spec version out in the field.

With the Vietnam War winding down, the defense market seemed shaky. Nevertheless, the idea found one taker. Jack Melchor, a private investor, guaranteed $100,000 in loans, which the four embellished with personal savings amounting to a grand total of $175,000.

ROLM began operations on June 1, 1969, in a prune-drying shed, an interesting variation on the garage theme. Coming up with the corporate moniker was the least of their problems, for it was simply an acronym referring to their surnames—*R* for *R*icheson; *O* for Dr. M. Kenneth *O*shman; *L* for Dr. Walter *L*oewenstern, Jr.; and *M* for Dr. Robert R. *M*axfield.

Once their blue-skying became a reality in the form of a tangible prototype, the quartet got an additional $600,000 from three venture-capital groups, one headed by oil-rich Saudi Arabian businessman Adnan Khashoggi.

After three successful years grinding out mil-spec computers, the ROLM founders once again grew restless. Their goal was to build a major American company, but they had the uneasy—and incorrect, as it turned out—suspicion that the market for their computer would be saturated by the time their annual sales reached $10 million to $20 million. Besides, they reasoned, the U.S. Navy was a major customer, and it looked like it was about to throw most of its business to Sperry Univac. It did. They decided to diversify.

The search for a second good product idea began, except this time the group was more systematic about it. They had one ground rule: Whatever product they chose, it had to be computer-based. After all, that was the embodiment of their expertise.

A gigantic feasibility study revealed that computerized telecommunications gear was the wave of the future and that stodgy, old, bureaucratic Ma Bell was vulnerable. Logically, they climbed aboard the PBX bandwagon.

Oshman points out, "We already had eighty percent of the technology. We figured we could get the other twenty percent easier than the telephone companies could get the technology."

They shipped their first 400-line PBX in 1975. They called it a CBX (short for *C*omputerized *B*usiness E*x*change). Three years later, ROLM's CBX, supplemented by software and other options to expand the system to accommodate hundreds more extensions, represented a major portion of the company's sales.

Ironically, the founders had misjudged the size of the mil-spec computer market. Today that branch of ROLM's business generates some $50 million annually, or about 15 percent of the firm's sales.

Still, the founders were right in steering ROLM into the telecommunications industry, where the company's future clearly lies. With their PBX products, ROLM and its competitors are placing their bets on the "office of the future"—an office fully automated and integrated. They're betting that voice and data transmission within an office will eventually converge into one digital "local area network." And the control center for this web of electronic neurons will be an advanced third-generation PBX.

ROLM's CBX II, which it introduced in 1984, is an example. The CBX II provides a very-high-speed digital network that connects all the desktop computers, minicomputers, and mainframes as well as telephones within an office or office complex.

Several of ROLM's competitors—including AT&T and newcomers InteCom, Inc., and CXC Corporation—hint they are also developing executive work stations through which tomorrow's executive will access services ranging from the facsimile transmission of documents to voice mail.

Such forward-thinking and long-range planning has made ROLM's four founders quite wealthy. By 1982, Oshman's 4.8 percent equity had a market value of about $29 million, for example, while Maxfield's 4.1 percent was worth $25 million. Richeson, on the other hand, owns much less because he retired from active money-grubbing back in 1975. He's a full-time volunteer with a religious and philosophical organization called Creative Initiatives.

Two years later, though, the value of the founders' equity more than doubled. In September 1984, IBM bought ROLM outright in an exchange of stock worth $1.25 billion.

Let There Be Light

The technology seems so rudimentary—not to mention obvious—in this age of invisible miracles. It was so obvious that Alexander Graham Bell demonstrated a type of telephone based on it as early as 1880.

Obvious or not, the technology of fiber optics took scientists a century to perfect. And now that they have, a whole new cosmos of marvelous communications possibilities is opening up to us.

Hairlike threads of the purest glass carrying beams of light are crucial to unlock the full potential of the information age. These fibers, known as "optical fibers," will provide the metaphorical superhighways for the transmission of all kinds of information—in the form of music, a television program, a telephone conversation, or the series of on-off pulses that animate computers.

Because optical fibers can transmit many times more information of the voice, data, and video varieties much more quickly than the old-fashioned copper wires or the conventional microwave transmission method, fiber optics will break the information bottleneck that's been holding us back. Fiber optics, in short, is the pivotal technology that will broaden the horizons of today's other emerging twenty-first-century technologies.

Alexander Graham Bell's photophone was based on the same principles as modern fiber optics technology—but what a crude rendition it was! To make his apparatus transmit voices, he used a thin, flexible mirror angled just right and plenty of direct sunlight. The photophone, clearly, wasn't practical. Still, Bell thought of it as his greatest achievement.

Bell also saw the photophone as the invention that would ultimately convince his detractors that he was more than just a flash-in-the-pan, a one-shot mechanical genius who had discovered the telephone by accident, as some Bell Telephone Company officials alleged.

Bell wrote to his wife, "I can't bear to hear that even my friends should think that I stumbled upon an invention [the telephone] and that there is no more good in me."

His wife encouraged him to ". . . bring out something no matter what, so it proves that was not the end of you."

The photophone was Bell's reply. Bell was so enthusiastic about it that he even proposed naming the couple's second daughter Photophone. Fortunately, common sense prevailed and she was named Marian.

In addition to being an unwieldy contraption the public did not understand, the photophone was soon eclipsed by Marchese Guglielmo Marconi's experiments with wireless telegraphy. The result was radio.

During World War I, there was a renewed burst of interest in lightwave communication in the United States and abroad. However, once the war ceased, so did the experiments. The United States government did not reinstitute formal lightwave research until 1941, when the lights were going out all over Europe once again.

Actually, the maturation and crossfertilization of four distinct technologies was necessary before contemporary fiber optics could be realized. They are the laser, invented in 1960; the solid-state electroluminescent diode or light-emitting diode called a "LED"; the solid-state photodector; and a low-loss optical fiber, which was introduced by Corning Glass in 1970. These technologies converged in the 1970s.

As accelerating developments in the field continue to drive down the costs, the commercialization of fiber optics is finally upon us in the 1980s.

Optical fiber systems are similar in appearance to the coaxial cable systems first installed by AT&T in the early 1940s, but the technology itself is quite different. In fiber optics systems, *light* is transmitted over a wispy-thin glass strand. In a coaxial cable system, *electricity* is transmitted over copper wire. This difference gives fiber optics tremendous advantages.

For one thing, fiber optics transmission has a larger bandwidth so it can carry vastly more information. For example, a single optical fiber as thin as a human hair can carry as many phone conversations as a bundle of the old copper wires as thick as a man's arm. This space-saving factor is important as more and more telecommunications cables are snaked through office buildings and city streets.

Lightwave communications systems also lower maintenance costs. For one thing, broken threads of glass in fiber optics cables are uncommon. Second, in a fiber optics system, repeaters, which amplify the signal, are spaced farther apart. This is because the transparency of optical fibers is such that light passing through them for distances of five hundred feet loses less intensity than it would passing through an ordinary windowpane. Fewer repeaters mean fewer devices that can fail.

But that's not all lightwave communication has going for it. Because optical fibers generally remain unaffected by moisture and temperature, these cables don't require pressurization as copper-wire cables do. Moreover, they're immune to the electrical interference that plagues the copper-wire-based system. That means no more static on the telephone line during lightning storms or annoying crosstalk due to induction; and less chance of eavesdropping by means of so-called wiretapping. The latter attribute makes it an excellent medium for military communication. Of perhaps more dubious utility, lightwave communication is even impervious to nuclear explosion.

Historically, one of the greatest barriers to fiber optics' progress was cost. The expense certainly had nothing to do with the price of the raw materials, since glass is about as abundant as sand. In contrast, copper is increasing in price because of its relative scarcity. No, fiber optics systems were high-priced for another reason: limited demand. The low-volume production created high unit costs. Today, as more telecommunicators and broadcasters convert to the newer technology, the unit price is dropping so fast one Merrill Lynch study predicts that during this decade, fiber optic networks will be about half the price of the conventional technologies.

The first small-scale use of a fiber optics cable for commercial transmission occurred in 1976, when TelePrompTer Manhattan Cable Television installed an eight-hundred-foot length at its headquarters. It carried signals from receiving equipment on the roof of a New York City skyscraper to a central processing center thirty-four floors below.

Commercial exploitation began in December 1980 when AT&T, after several years of trials, installed fiber optics cable in part of its system in Atlanta. Western Electric was the supplier. GTE, MCI, and several other big potential users immediately announced their intention to follow suit.

Thus the AT&T installation functioned as a beacon to budding entrepreneurs. The message was: "There are hungry buyers out there who will let substantial contracts to those suppliers who have something extraordinary to sell."

Suddenly Corning Glass and Bell Labs, the incubators of lightwave technology, found their best engineers deserting to set up their own production facilities. The opportunities in fiber optics were plentiful enough that many of the early entrepreneurs wisely staked out different segments of the nascent industry. Plentiful, too, was the investment capital to fuel the industry's growth.

In Signal Hill, California, for example, E. Blaine Mansfield, Jr., and W. Edward Naugler, Jr., teamed up in 1980 with an ex-Bell Labs scientist named Hargovind N. Vazirani to found American Fiber Optics Corporation (AMFOX). They went public a year later, even though AMFOX's main technology was based on an as-yet-ungranted license from Western Electric. Still, they managed to raise $4.5 million.

Hargovind Vazirani has since left AMFOX, but Naugler and Mansfield each retain between 15 and 20 percent equity. However, their company faces an uphill battle, since the heavyweights in the industry—Corning, Western Electric, ITT, and Valtec—seem determined to maintain their leads in the production of the delicate optical fibers.

Also based in the Los Angeles area, Lightwave Technologies, Inc., was started in 1981 by another ex-Bell Labs engineer, Dr. Frank W. Dabby, with the assistance of Jim Goell, hailing from ITT's fiber division, and a third, silent partner. The trio managed to retain 100 percent equity in the company by securing a seven-year $1 million capital equipment lease from a Bank of America subsidiary. But when the company went public in 1984, Dabby and his mother emerged as by far the largest shareholders with 26 percent.

Lightwave Technologies is actually Dr. Dabby's second company. In 1975 he founded Fiber Communications, which still makes the more common "graded index" fiber. He sold it two years later for $5.6 million. His goal for Lightwave Technologies is to make it a leading producer of single or "mono mode" optical fiber, which many think will eventually become the industry standard.

On the East Coast, where Bell Labs and Corning are headquartered, the start-up activity is even more feverish:

Richard Cerny's (Corning alumnus) and Tadeusz Witkowicz's Artel Communications Corporation in Worcester, Massachusetts, has been growing fast since its 1981 founding. It makes fiber optics transmission systems used for applications ranging from teleconferencing and broadcasting to data transmission.

Not far away in Sturbridge, Dr. Raymond E. Jaeger (ex-Bell employee) and partner Dr. Mohd A. Aslami (ex-Corning) raised $1.5 million in a private stock placement in 1981 to finance SpecTran Corporation's fiber production facilities. Two years later, in 1983, SpecTran became one of the few fiber optics start-ups to go public. Jaeger and Aslami each held 18.1 percent equity, which dropped to 11.6 percent after the offering.

Norrsken Corporation, a little to the southwest in Cheshire, Connecticut, is the brainchild of Dr. Eric N. Randall (Corning alum). Dr. Randall founded it in 1981 to build fiber optics manufacturing equipment for the big guns of the industry as well as for the industry's aspiring entrepreneurs.

Dr. Randall, a physical chemist, has the rare distinction of being one of those few who were there at the genesis when Corning's chemistry research department achieved the world's first optical fiber suitable for telecommunications applications.

And then there's the company bearing engineer-journalist Irwin Math's name in Port Washington, Long Island. Started in 1977 as a mail-order operation for fiber optics components, Math Associates has since become a full-fledged manufacturer of fiber optics data transmission systems.

PART FOUR

The Stay-at-Home Society

7 Moving Pictures: *The Electronic Entertainers*

The spirit of Walt Disney lives on in Sunnyvale, California, in the body of a bearded young entertainment tycoon. During the 1970s, this innovative engineer used computer-graphics technology to revolutionize America's definition of leisure. Over the next two decades he promises to push the country's recreational activities even farther into the twenty-first century. True, this young man's personal life might not bear the scrutiny of Walt Disney's, but his genius certainly runs along lines equally creative, if not more technologically sophisticated.

His name is Nolan K. Bushnell, ex-Mormon, once divorced, remarried with a second family. He's the man largely responsible for the popularity of arcade video games.

While Nolan Bushnell and his business counterparts are hardly the corrupters of youth some parental vigilante groups painted them to be during the height of the video games craze, Bushnell is no purveyor of wholesome 1950s-style family fun either. Oh, he's all for family fun. Why not? It once put oodles of quarters in the coin slots of the arcade games that lined the walls of his robot-infested Pizza Time Theatre parlors. But Bushnell also recognizes that times have changed, and his sense of fun and humor reflect those changes. It's the cloying word "wholesome" that doesn't fit.

What's changed?

Mr. Bushnell and his adolescent fans in the 1970s and 1980s live in the shadow of a controversial Southeast Asian war that America lost, in a time when the country is lowering its expectations and retrenching. Conversely, Mr. Disney and his fans basked in the glory of a worldwide war that America won, the aftermath of which were rising expectations and booming prosperity. Walt Disney captured this sense of national pride and optimism in the guise of a do-gooder mouse named Mickey and a gadabout duck named Donald. In contrast, Nolan Bushnell's galaxy of animated robot characters is altogether more ethnic and decidedly less inclined toward motherhood and apple pie.

If you think war and the military don't have much to do with contemporary video games and the other computer-based forms of rest and relaxation we're about to discuss, think again. The military is the source for much of this new blip-and-zappo entertainment technology.

"The joke in the electronics industry is that the technology in video games [in the early 1980s] is more advanced and more militarily useful than some technology banned for export by the Department of Commerce," says Ken Bosomworth, the president of International Resource Development, a market research firm that tracks the electronic games business. "The people creating games are using 1985 hardware and the United States military is using 1965 hardware."

Apparently the American military is aware of this technology gap and now seems on full alert to the benefits it can derive from electronic gaming. More and more of the U.S. armed services' combat simulations resemble the best games that American industry has to offer. At Fort Eustis in Virginia recruits are trained with a version of Atari's "Battlezone," whose targets are realistic silhouettes of enemy tanks, helicopters, and armored personnel carriers. And consultants from Atari helped the United States Army create a tabletop tank gunnery game, the MK-60. It costs $15,000 and comes with thirty complex programs.

Some military masterminds even appear cognizant of the human factor—that the video arcadians of today, with their superior spatial sensibility and hand-eye coordination, will make excellent soldiers of tomorrow. To attract new recruits, Major Jack Thorpe of the Pentagon's Advanced Research Projects Agency opines: "It's important to have training devices that don't appear so obviously to be training devices."

Thus it should come as no surprise that Sanders Associates of Nashua, New Hampshire, a respected contractor for the American military, designed one of the early forerunners of the video game. Actually, it was all quite serendipitous.

For ten years, Sanders's employee Ralph Baer had been working on radar and electronic defense systems. Then one day in 1967 for no particular reason, this German émigré engineer began tinkering with a stray television set someone had left around the lab. Before long, he had a couple of blobs chasing each other around the screen. The video games concept evolved from there. With a little more fine-tuning and some refinements, Baer and the other engineers who were eventually assigned to the project had a device that could be configured to play back-and-forth reaction games like hockey and Ping-Pong.

Sanders Associates patented the device and waited. Finally, some three years later, Magnavox expressed interest in licensing the technology.

In 1972, Odyssey, the first primitive home video game, hit the market. Sporting a $100 price tag, its mass-consumer appeal was understandably limited. Nevertheless, nearly a hundred thousand Odysseys were sold the first season.

The next time out, Ralph Baer was smarter. He devised his follow-up product on his own time and licensed it through Marvin Glass & Associates. The invention was Simon, as in "very simple," a kiddie toy whose only link to latter-day video games is its electronic innards.

Simon is a disc divided into sections, with buttons in the middle. Each section, when pushed down, emits a different pitch and glows a different color. The object is to do what Simon says: remember and duplicate a sequence of pitches that grow ever longer as the game progresses.

Simon was Milton Bradley's entry in the electronic toy carnival in 1978. Its opposite number was an equally noisy microprocessor-based miracle that looked like a modernistic telephone and answered to the name Merlin. Merlin arrived compliments of Parker Brothers (Milton Bradley's arch rival) and Bob and Holly Doyle, the husband-and-wife team of astrophysicists who invented it in their unassuming, three-decker house in a middle-class section of Cambridge, Massachusetts.

Stripped down to basics, Merlin is nothing more than a hand-held computer with a memory infallible enough to play a small library of conventional games (e.g., tic-tac-toe, blackjack, magic squares) and drone out tunes of the player's own devising.

Steve Hansen

Electronic toys were Robert and Holly Doyle's bailiwick until they invented a palm-sized computer terminal that communicates with data banks over the telephone. This husband-and-wife team of astrophysicists works out of this third-floor office in their Cambridge, Massachusetts, home.

By 1983, Merlin's magic had spirited almost $200 million from parental pocketbooks both here and abroad, and it has showered the Doyles with more than $6 million in royalties. The Doyles are using this grubstake to update their wares for information-age consumption. Their latest product, manufactured and marketed by their own firm, called IXO, Inc., is a toy-size computer terminal that communicates with data banks over telephone lines. Unfortunately, it has yet to seize the imaginations of adults to the same degree that Merlin and their other electronic toys seized those of children.

* * *

In 1972, the same year Odyssey introduced game-playing Americans to the joy of high-tech rec, Nolan Bushnell and a partner founded the company that is really responsible for detonating the video games explosion.

Bushnell's early exposure to computer graphics, the technology underlying video games, occurred in the 1960s, when he was studying electrical engineering at the University of Utah. The school had an excellent computer science department headed by Dr. David C. Evans, a pioneer in the computer graphics field. Evans subsequently teamed up with Ivan E. Sutherland, a younger colleague in the electrical engineering department, to found the Evans & Sutherland Computer Corporation, a company that's become a leading producer of computer graphics systems for an eclectic group of customers ranging from automobile designers to special-effects artists and animators in the film industry.

It's an amusing footnote that Evans & Sutherland's first product was an extraordinary piece of hardware called the LDS-1, which purportedly stood for "*L*ine *D*rawing *S*ystem." But to the citizens of Salt Lake City where the company is based, LDS stands for only one thing: Latter-Day Saint. Dr. Evans is one.

Evans & Sutherland

Dr. David C. Evans (left) and Ivan E. Sutherland teamed up in 1968 to found Evans & Sutherland, one of the pioneer firms in the emerging computer graphics field. Today Dr. Evans is the only founder still involved in the company's management.

At universities during the free-speech sixties, the scenario was the same: Liberal-arts students occupied themselves with protests against "the establishment's war in Southeast Asia" while Bushnell and his fellow techies in the engineering department squared off against their schools' respective mainframe computers every night in a game called Space Wars. Space Wars was a sophisticated—at least for its time—game program written by an MIT graduate student named Steve Russell. It was the underground classic that initiated a whole generation of hackers in the delights of computerized intergalactic mayhem. Needless to say, Space Wars cost many large companies and universities millions of dollars in computer time until commercial video games finally made their appearance in the 1970s.

Every summer during Bushnell's college years, he worked as a sideshow barker and manager of the games arcade in a local amusement park. The attractions were the old standbys—knocking over milk bottles with a baseball or shooting down moving duck decoys to win a stuffed kangaroo. Nolan Bushnell refers to his years on the midway as "an intensive behavioral training session" that taught him what people respond to in the way of recreation. He thought they'd respond to a computerized game like Space Wars just fine, but since the technology for a miniaturized, coin-operated version wasn't available yet, he shelved the notion for future reference.

Bushnell graduated in 1968, about the time minicomputers were coming into their own. He got a job as a research engineer working for Ampex Corporation in Redwood City, California, and watched with astonishment as the price of computing power plummeted from $40,000 to $4,000 over a two-year period. The time, he decided, was ripe for developing his computer-animated game idea. With the invention of the microprocessor in 1971, he knew it was now or never.

It was an arduous gestation period. Moonlighting from his $12,000-a-year job, Bushnell would return home to his tract house in Santa Clara each evening and hole up in his daughter's bedroom, which he'd converted into a lab. After months of such antisocial behavior, which culminated in a divorce several years later, Bushnell created an adaptation of Space Wars called Computer Space.

Computer Space was a coin-operated arcade game that pitted spaceships against flying saucers. It was probably too sophisticated for the average layman—but Bushnell's engineer friends loved it.

By this time Bushnell had quit Ampex and teamed up with a

neighbor, Joseph F. Keenan, an IBM computer salesman, in a partnership called Syzygy. They sold the rights to Computer Space to Nutting Associates, and Bushnell went on its payroll to oversee the game's production.

About two thousand Computer Space units were sold, certainly no blockbuster but enough to whet Nutting's appetite for Bushnell's next offering. When Nutting's management broached the subject, Bushnell countered with a demand: In exchange for his next wrist bender, he wanted an option on a third of the company's equity and more say about marketing strategy. Nutting made a counteroffer: "Engineers should stick to engineering," they said. "However, since you're a talented one, we'll give you an option on five percent equity."

Bushnell marched. Within days, he and Keenan had incorporated Atari because they'd discovered the name "Syzygy" was already taken.

The royalties from Computer Space underwrote Pong, the game Atari introduced in 1972. Pong made Nolan Bushnell famous even though he wasn't, as legend has it, its sole author.

According to Bushnell, Pong was "a mistake. . . . I had just hired a new engineer named Al Alcorn and I asked him to write a program for a simple Ping-Pong kind of game. I told him how the ball should act and described the sound it should make. It was a throwaway, really, just an exercise to get Al up to speed for a driving game that I really had in mind. Pong turned out to be a lot of fun, so we decided to market it."

To test it out, Bushnell and Keenan placed a prototype in Andy Capp's Tavern in Sunnyvale. A couple of days later, they got an emergency call reporting a broken machine. A thorough inspection revealed a simple explanation: The coin box was overstuffed. Pong was gagging on quarters. It was prophetic. Before the Pong fad peaked, an estimated one hundred thousand units were sold—and only 10 percent of them by Atari. The game was so popular that counterfeits, many of them Japanese-made, were rife.

Pong unilaterally ignited the video games phenomenon and relegated that pool-hall mainstay, the mechanical pinball machine, to the scrap heap.

Over the next few years, Atari developed and marketed some thirty-five other arcade games with only middling success—that is, until the 1976 debut of Breakout. Breakout requires players to knock a hole through solid brick walls to escape from prison. It sold fifteen thousand and convinced Warner Communications that Atari, Inc., was worth the $28 million it paid for the company that year. MCA, Disney, General Electric, and Columbia Pictures each had considered the purchase and declined.

Bushnell and Keenan agreed to the sale because it would help them underwrite their long-term, mass-market strategy. The duo had made one abortive pass at the home consumer already. For the 1975 Christmas season, they'd launched a home-game version of Pong that had sold out before it reached the stores. In addition to piquing their interest, the whirlwind experience had given them a glimpse of the problems that lay ahead. It taught them that the transition from the self-financing, year-round arcade market to the home market, which relies heavily on the once-a-year largesse of a mythical character in a red suit, required massive infusions of capital they didn't have. And neither one of them wanted to dilute their equity to get it.

Before the sale to Warner Communications was finalized, Bushnell, an Edwardian-looking man with the Sybaritic tastes to match, committed an indiscretion that nearly proved fatal. Nolan Bushnell, the *nouveau* millionaire, made the mistake of letting a local newspaper print a photograph of him enjoying the good life with a young lady friend in a hot tub. His ex-wife reacted with a lawsuit, raising questions about whether Bushnell held clear title to his Atari shares, representing over 50 percent of the company's equity. Warner's high-powered attorneys intervened and induced the ex-Mrs. Bushnell to settle for a minuscule sum. As a result, Nolan Bushnell realized $15 million from the sale and took up the post of a Warner division chieftain.

Warner injected its new subsidiary with cash and waited for the bonanza. Atari's first innovation was a more versatile type of home video game introduced in 1977. Instead of the "dedicated" (one-game-only) electronic device, Atari introduced the "programmable" video game. Atari dubbed it a VCS, short for "*v*ideo *c*omputer *s*ystem." The VCS and the me-too systems to follow from other manufacturers came in two parts. The hardware consisted of a "module," a compact rectangular box with a long cord that plugs into a television set. The software was a game "cartridge" that you insert in the module. This insertable cartridge meant you could change game programs as easily as you change the cassette in a tape recorder or the records on a phonograph.

The bonanza finally struck Atari like a gusher when it licensed an arcade game, Space Invaders, to sell to the living-room set in 1980. Zoooooooooommmmm went Atari's sales. Its revenues doubled and its operating profits quintupled within the span of a year, making the company the indisputable industry leader. Needless to say, this multibillion-dollar blast immediately attracted the mightiest of competitors in both the arcade and home segments of the video game machine marketplace.

But for the purposes of this book, the arcade and home-hardware segments of the video games industry have one unfortunate feature in common—an almost universal absence of entrepreneurs. The reason was a matter of timing and economics. When the electronic games industry finally took off, only those established companies technologically prepared, with money already in the bank, were in a position to exploit this rapid-fire marketplace. Predictably, the companies that jumped into the fray were those already in related businesses—casinos and amusement parks, toys, pinball machines, and jukeboxes.

But if video games hardware held no allure for the would-be entrepreneur, this couldn't be said about the more inviting software side of the business. Software has proved a treasure trove for a select few game designers with an entrepreneurial hankering who got into the business early enough. While the investment needed to start a games hardware company is substantial, it requires relatively little to develop and manufacture game cartridges and disks.

Activision, Inc., founded in the fall of 1979, was the first independent company to produce cartridges for Atari's VCS. Until then Atari had been the sole producer of games for its machine.

If you accept the premise that software sells hardware, you'd expect Atari would have welcomed Activision into the arena. It didn't. Atari reacted with a $20 million lawsuit charging Activision's founders with unfair competition and conspiracy to steal trade secrets, a suit eventually settled out of court.

Atari's consternation was largely due to the loss of four of its most talented game designers—Alan Miller, David Crane, Bob Whitehead, and Larry Kaplan. It was these well-educated young men, all in their late twenties, who teamed up with a veteran marketing executive, James H. Levy, to start Activision. The group secured almost $1 million in venture capital from Sutter Hill Ventures, which became the company's majority shareholder.

The stifling corporate milieu at Atari, the reason for the foursome's disenchantment, was not replicated at Activision's Mountain View, California, headquarters. Anything but. Unlike Atari, Activision coddles its designers, a smart move since the pool is limited, comprised of perhaps as few as a hundred topflight "artists" who possess the requisite combination of technical programming skill and creative genius. Indeed, video game designers are to the 1980s what rock stars were to the 1960s and 1970s. Activision's designers set their own hours, conceive their own game projects, get top billing as author, and stand to profit handsomely from royalties—in addition to a substantial base salary—should one of their games hit the jackpot. Among other distinctions, Activision employs one of the few women in the field, designer Carol Shaw.

Another more recent Silicon Valley game producer, incorporated in 1981, seems a mirror image of Activision. It, too, was greeted by a lawsuit from Atari, since settled. It, too, was founded by ex-employees from Atari (William F. X. Grubb and Dennis Koble) augmented by two defectors from Mattel Electronics (Brian Dougherty and Jim Goldberger). It, too, had venture-capital backing—but double that of Activision and from the firm of Kleiner, Perkins, Caufield & Byers. It, too, produced cartridges for both Atari's VCS and Mattel's Intellivision console. And it, too, had sales in its first year exceeding $50 million.

A showdown with Warner Communications' management over the future direction of Atari had provoked Bushnell's resignation in January 1979. One of the precipitating issues involved Atari's lack of support for a Bushnell pet project: the mingling of robot characters à la Disneyland with coin-operated arcade games and fast-food fare in a chain of family entertainment restaurants. Bushnell called this zany combo Pizza Time Theatre. Bushnell had persuaded Atari to open a prototype restaurant in May 1977 in San Jose, but nothing further was done. After his departure, Bushnell bought back all rights to the concept for $500,000 and, with the aid of Joe Keenan, proceeded to develop it.

The master of ceremonies at all Pizza Time Theatre restaurants is Chuck E. Cheese, a tough-talking rat who sounds like he ought to be driving a New York City taxicab. None of the other robotic characters, suspended from boxes along the walls of the dining area, lacks for chutzpah, either. The mix of characters in each restaurant varies, but among the most popular of the fifteen in Bushnell's troop are Jasper T. Jowls, a floppy-eared dog who strums a banjo; Pasquale, a moustachioed Italian chef who breaks into operatic song until Mr. Cheese shuts him up; Madame Oink, a Parisian pig reminiscent of Miss Piggy; and Harmony Howlette, a country-and-western warbler patterned after Dolly Parton.

But computer-driven robots aren't all Pizza Time has to offer.

Pizza Time Theatre

Nolan Bushnell, the 1980s' answer to Walt Disney, is pictured with Chuck E. Cheese, the robotic character who acts as master of ceremonies at Bushnell's chain of Pizza Time Theatre restaurants. Bushnell's first venture was Atari, a video games company he sold to Warner Communications.

After the kiddies are fed and tire of these characters' intermittent antics, they can wander off into an adjoining video games arcade.

Bushnell's formula worked. Parents seem particularly pleased that their children at last had an acceptable place to hang out while they vented their videologue fantasies. In fact, the initial popularity of the arcade games made earth-moving equipment seem attractive to some Pizza Time proprietors as a means to haul away the mountains of quarters they'd collected.

Needless to say, each piece of that silver coinage added one more increment to the windfall Nolan Bushnell eventually realized when his franchised chain went public in 1981. Bushnell's 33.5 percent equity was valued at close to $35 million. The share of his second wife, Nancy, was worth almost $2 million, and Joe Keenan's 4.3 percent, some $5 million.

A tribute to Bushnell's ingenuity is the train of copycat franchisers Pizza Time inspired. The only one given much hope for long-term success is ShowBiz Pizza Place, the construct of Robert Brock.

Robert Brock made his first millions in the southwestern corner of the United States as a franchisee of Holiday Inns. His Brock Hotel Corporation originally worked out a deal with Bushnell to build as many as two hundred Pizza Time restaurants in the Midwest and South. But their arrangement dissolved into a bitter lawsuit after Brock discovered he had another option that would allow him to develop his own competing chain. Eventually the suit was settled when Brock agreed to pay what amounts to a licensing fee over a period of years for use of Pizza Time's concept.

The option that caused Brock all the trouble presented itself in the form of Aaron Fechter, a young Florida inventor who had been making computerized robots—many think better ones than Bushnell's—for amusement parks since 1977. Fechter's company, Creative Engineering, in Orlando, is now Brock's partner.

Creative Engineering supplies ShowBiz with Billy Bob Brockali and the rest of its restless menagerie in exchange for 20 percent ownership. Creative Engineering is, in turn, owned 71 percent by Aaron and 20 percent by his parents, who put up the initial $5,000 investment to start the company. Moreover, the contract affords Creative Engineering the freedom to continue providing electronic entertainment for theme parks and any other type of recreation that doesn't compete directly with ShowBiz Pizza.

Creative Engineering

Aaron Fechter is the young Florida inventor who bartered his expertise at making robotic characters for 20 percent ownership in ShowBiz Pizza Place, the chain that competes with Nolan Bushnell's Pizza Time Theatre.

In 1983 it happened. What the video games industry predicted would happen toward the end of the decade struck five years too soon. By the end of 1983, the devastation was complete.

The makers of video game consoles and cartridges had lost collectively more money in that one year than they'd made collectively in the good years. As an industry, they'd lost more than $1 billion. Financially the industry was experiencing the same phenomenon it had peddled so successfully for three years—chaos, death, and destruction. As a consequence, economic observers were relegating dedicated video games paraphernalia to the same corner of the attic inhabited by broken-down hula hoops, Pet Rocks, citizens band radios, and Rubik's Cubes.

Nor was Bushnell's Pizza Time Theatre, with the added attraction of its singing robots and fast food, cushioned from the impact. In March 1984, the chain succumbed to Chapter 11. In Pizza Time's case, teenage boredom with video arcade games wasn't the sole reason for its problems. There was also an overly aggressive expansion campaign that resulted in more than 250 units, some practically across the street from each other. And there was its high price for mediocre "cardboard" pizza.

True to his instincts, Bushnell had sensed the public's growing fatigue with his entertainment formula several years earlier and moved on to more adult ventures. Almost immediately after it opened in the spring of 1982, his Lion and Compass restaurant in Sunnyvale became the mecca for Silicon Valley dealmakers in the same way that the Polo Lounge is for Los Angeles movie-industry moguls. In price, quality, clientele, and ambiance, Lion and Compass is the polar opposite of a Pizza Time eatery. It's the epitome of laid-back California chic.

Bushnell also owns Catalyst Technologies, a holding company for nascent high-tech companies, and Androbot, Inc., which manufactures an embryonic line of home helpers operated as peripherals to microcomputers and video game consoles. Granted, Androbot's first androids—sporting such names as Topo, FRED (for *F*riendly *R*obotic *E*ducational *D*evice), and BOB (for *B*rains *o*n *B*oard)—have about as much intelligence as your pet gerbil. You could, however, regard them as futuristic pets—or, viewed from another angle, as toys for eccentric multimillionaires like Nolan Bushnell.

But despite the red ink and bankruptcies, video game playing per se is not an extinct form of recreation. True, it's passed its apogee as a fad, but in the process, it entered the culture, taking its place beside such other leisure-time alternatives as moviegoing, TV watching, and sports spectating. The difference is that, henceforth, more and more video games will enter American households on floppy diskettes that family members insert into the disk drives of their trusty, multipurpose home microcomputers. Games on computer disks are replacing yesterday's cartridge software.

This trend is now well under way. The ubiquitous Apple microcomputers, for example, attract many buyers largely because of their extensive library of game programs available from a galaxy of small software companies, mostly California-based.

Sierra On-Line, Inc., came about more by accident than design.

Kenneth Williams, a programmer for a Los Angeles computer consulting firm, arrived home one evening to find his wife, Roberta, in the act of creating a complex whodunit complete with spooky mansion, stolen gems, and roving murderer. The idea occurred to them that the story could be translated into some sort of computer game, albeit not your usual shoot-'em-up type. Eventually the couple managed to turn it into a viable program, which they sold through computer magazine ads under the title Mystery House.

Their initial investment of $1,200 returned a gross of $167,000 that first year of 1980. Ken Williams quit his job, and the Williams family, which includes two children, took the money and moved to more scenic climes. In the tiny town of Coursegold, seated in the Sierra Nevadas on the fringe of Yosemite National Park, they set up housekeeping and shopkeeping.

To underwrite this idyllic life-style, Roberta Williams set about

Fresno Bee

Kenneth Williams was a programmer for a Los Angeles consulting firm when he arrived home one evening to find his wife, Roberta, writing a whodunit. Ken turned Roberta's mystery into a dandy microcomputer adventure game. That and other games gave the Williamses the capital to move north to the Sierra Nevadas and found Sierra On-Line, Inc.

spinning more fantastic yarns, the second of which was The Wizard and the Princess, an even bigger seller. And Ken Williams began recruiting other programmers whose diverse talents would help broaden the product line of Sierra On-Line, Inc.

"Diverse" is a good word to describe the company's ninety-odd software programs. They include a growing series of adventure games, a hit arcade-style game called Frogger, a word-processing program, and an adults-only item dubbed Softporn in which male players seek to seduce three women with all the attendant hazards that entails. The Softporn packaging displays Roberta Williams and two other nubile maidens barely breasting the waves in a California hot tub.

Lest you think this nonsense, it might be well to point out that TA Associates, a respected Boston venture-capital firm, took Sierra On-Line's version of fun and games quite seriously. TA found its repertoire worthy enough to underwrite $3 million more of the same.

Battles—actual campaigns drawn from historical records—are the domain of Joel Billings's firm, aptly named Strategic Simulations, Inc. (SSI). It's the publisher of such complex military masterpieces as Computer Bismarck, The Campaigns of Napoleon, Guadalcanal Campaign, and Pursuit of the *Graf Spee.* Billings's offerings attract the hard-core history buff and intellectual, people who are motivated to devote anywhere from six to sixty hours to these games of strategy with their infinite number of possible outcomes.

Because SSI's early games did not attract the mass of videonauts who frequent arcades, Billings saw the wisdom of branching out into fantasy war games. Strategic Simulations' The Warp Factor is a hypothetical space war game set in the future instead of the past. But it's still a tactical game emphasizing solid decision-making, since the players choose their weapons and battle sites.

An ex-Air Force lieutenant colonel is actually filling the corporate cockpit of one Sacramento, California, software house. Terry Bradley started Sirius Software in 1980 with the proceeds from the sale of his ComputerLand franchise and piloted the company's successful launch. It's been plagued, however, by a malady common among game producers—key defections from among the ranks of both its in-house and free-lance design staff.

Nasir Gebelli, an Iranian who came to the United States to study computer science and stayed to write best-selling programs such as

Gorgon, certainly won't be licensing any more of his creations to Sirius. Instead, they'll be the property of his own firm, Gebelli Software, which is located nearby.

Spinnaker Software, Inc., and MUSE Software, two East Coast entries, also have designs on the home Apple computer owner.

Spinnaker's cofounders, William Bowman and David Seuss (pronounced "cease"), met at Harvard Business School. They both worked for Boston Consulting Group before teaming up and receiving funding from TA Associates in 1981. Their firm's thrust is educational, but their programs resemble video games more than classroom exercises, attempting to amuse children while teaching them something.

MUSE's products range from educational and word-processing software to video games. This Baltimore company began as the moonlighting endeavor of a computer analyst named Ed Zaron. Tank War, ESCAPE!, and The Maze Game—his first three games designed with the help of a friend, Silas Warner—brought in enough money to warrant incorporating in 1978. Since its inception, MUSE has been completely self-financed.

But it's a California denizen who is, perhaps, the most legendary figure in the Apple computer cosmos.

With his strapping build, boyish good looks, and literary inclinations, Bill Budge belies the image of the solitary computer nerd. Before founding his BudgeCo., he actually managed to spend lengthy periods of time without touching a computer keyboard. Instead he concentrated on such nontechnical endeavors as short-story writing, for he originally aspired to be a novelist. This may account for his breakthrough approach to computer-games graphics.

Budge's 1980 game Death Star represented the first use of genuine three-dimensional effects in a microcomputer program. His big winner to date is Raster Blaster, a title inspired by Stevie Wonder's song "Master Blaster." Fans say what differentiates this computerized pinball game from all the others is its verisimilitude. It looks and feels like a real pinball game.

Miraculously, Budge achieved his closer-to-reality effects without resorting to several of the emerging technologies guaranteed to create more lifelike images and sounds. These new technologies represent the wave of the future in computerized entertainment. Among them are *holography,* a type of photography whereby a beam of laser light constructs three-dimensional images that seem to float magically in space . . . *speech synthesis,* or games that talk . . . and *voice activation,* computerized games that do your spoken bidding.

But in terms of pure spectacle, the *interactive videodisk* is probably the most exciting new technology and the one that entertainment pioneers like Bushnell are likely to exploit first.

Videodisk technology represents the symbiosis of computer- and laser-generated sound and imagery. As a game and educational medium, it has everything going for it. For example, using videodisk techniques, for the first time showers of astral garbage on a videoscreen will look and sound like streaking trash, replacing yesterday's unimaginative cartoon blips. This is possible because videodisks utilize high-resolution visuals comparable to film footage. And a videodisk's interactive capabilities mean a player can stop the action anywhere he pleases, just as filmmakers use freeze frames. If a player wants to jump forward or backward, randomly accessing information, that's possible, too.

Nowhere are these features better demonstrated than in an experimental videodisk created by the Architecture Machine Group at MIT. On the disk, the MIT group constructed a visual record of every street and building in the small mountain town of Aspen, Colorado, during the four seasons of the year. The moving images appear on the screen from the perspective of a person driving or walking down the middle of those streets. As the viewer, you can use a computer keyboard to indicate where you want to go in Aspen at what time of the year. It's winter and you want to get out of the ice and snow and enter the courthouse. Press the appropriate cursor key and you'll find yourself walking up the steps into the building and seeing what any real person would see had he made the same decision.

MIT's Aspen "movie map" rates very high on the gee-whiz scale. It proves that the potential of this medium as an educational, entertainment, and even computer-storage tool has hardly been scratched.

That potential has not been lost on George Lucas, the late-thirtyish professional film editor/screenwriter/director/producer and industry visionary. As the founder and principal stockholder of Lucasfilm Ltd., the bearded Mr. Lucas has become over the past few years a major force in the motion-picture business, a man who has set himself the idealistic (perhaps impossible) goal of regenerating and recasting the whole industry. Should he prevail, the film community's locus, leadership, subject matter, and production methods in the future will be a far cry from anything it's known in the past. Indeed, one of Lucas's major goals is to use the most sophisticated computer-graphics equipment available to increase the number of eye-catching pyrotechnics in movies as well as video games while simultaneously reducing the number of man-hours that go into their production.

George Lucas, you'll recall, is the young man who visited that dizzying wonderland of technical fireworks known as *Star Wars* on an awestruck moviegoing public in 1977. From every possible angle, the film was an epic: It told an involving and ageless story that would have done the ancient Greek mythologists proud. It showcased special effects that appeared to be computerized and state-of-the-art although they were, in actuality, nothing more than extraspectacular examples of traditional mechanical movie wizardry. And perhaps most important of all, it provided Lucas, the middle-class boy wonder from the hot and dusty mid-California farm town of Modesto, with the grubstake he needed to implement his anti-establishment ideas.

Star Wars put at least $25 million into Lucas's pocket, not to mention the millions it filtered into the bank vault of Lucasfilm, his corporate alter ego. To be accurate, though, Lucas was not the only individual to profit handsomely. He insisted on dispersing a sizable portion of the *Star Wars* proceeds, in the form of generous bonuses, among the film's crew and its three lead actors.

The *Star Wars* sequels—*The Empire Strikes Back* and *Return of the Jedi*—were also extraordinarily profitable. This trilogy and the licensing income from spin-off products gave George Lucas a net personal worth approaching $100 million and the wherewithal to build at last his infotainment utopia on a hilly blanket of land north of San Francisco.

Skywalker Ranch is a Lucas fantasy transformed into reality, the stuff out of which cinematic dreams are woven. On his three thousand acres, Lucas is erecting what he likes to call a "backyard film studio." Upon its completion, Skywalker Ranch will give Lucas's philosophical compatriots access to the most advanced automated equipment for constructing movies ranging from epic to minuscule proportions. But to Lucas's way of thinking, the size of a film's

Wide World Photos

George Lucas is the crusading boy wonder who turned his mythological ideas into the *Star Wars* trilogy, movies that sent Hollywood in pursuit of films showcasing state-of-the-art special effects. Meanwhile, Lucas took his newfound fortune and retreated to Marin County, where he's building a movie empire around twenty-first-century technology.

canvas misses the point. The heart of the issue is the use of twenty-first-century technology to tell stories, large or small, manifesting a back-to-basics counterculture social vision that is, in many respects, closer in feeling to the previous century than the current one.

If George Lucas's vision is realized, aspiring filmmakers graduating from cinema schools will fan out across the country, making movies that reflect the grass-roots outlook of the heartland. And eventually, he hopes, their world view will replace the stale and narrow vision of contemporary entertainment-industry kingpins who exercise dominion over the American scene from their East Coast and West Coast strongholds. According to Lucas, their power is out of all proportion with their knowledge of what's really happening in the minds of real people.

Star Wars represented more than just a personal triumph for George Lucas. It epitomized a turning point in the history of the American film industry.

To the old-style Hollywood dream merchants whom Lucas is trying to thwart, the message of *Star Wars* was clear: The public wants more and better special effects. And what the public will pay for, Hollywood always tries to deliver even if it means moving into technological realms heretofore unexplored by the denizens of tinsel town.

Naturally, it's George Lucas and his like-thinking counterparts in the industry who are leading the cinematic search party into that electronic netherworld where angels previously feared to tread. Pushing out the boundaries of filmmaking innovation are such Lucas pals as Steven Spielberg *(Jaws, Close Encounters of the Third Kind, Raiders of the Lost Ark, Poltergeist,* and *E.T.),* Francis Ford Coppola *(The Godfather* and *Apocalypse Now),* and Steven Lisberger *(TRON).*

TRON, released in 1982, is the first feature film to rely heavily on a computer-simulated environment to advance its plot, which revolves around a determined video games designer and his odyssey inside a computer. Ironically, it was produced by Walt Disney Productions, a studio with its roots in the Hollywood of yore.

When a movie has sequences that combine computer-generated imagery and live action (that is, real actors placed in settings, even costumes, created by a computer), there are only a handful of computer animation houses where a producer can subcontract out the job. Disney sought out several of them to create the computer-generated film footage for *TRON.* Working independently of each other, these companies originated the sixteen minutes of *TRON* that consist entirely of computer-animated action and the forty-five minutes that employ the 150 different computer-drawn backgrounds and sets in front of which the film's stars appear to be performing.

Mathematical Applications Group, Inc. (MAGI), tucked away in a science park in Elmsford, New York, is responsible for the bedazzling motorcycle race in *TRON.* Its credits also include several medical and scientific films using computer animation to fabricate that microcosmic universe the naked eye will never see.

MAGI was one of the earliest firms to turn a profit, albeit a small one, from computer graphics products and services. It was incorporated in 1967 by Dr. Phillip S. Mittelman, a Harvard- and Rensselaer-educated nuclear physicist.

Dr. Mittelman was employed by United Nuclear Corporation when he first stumbled into the colorful world of computer modeling.

Mittelman managed a group of engineers who had developed a program depicting how radiation beams would behave inside variously constructed atomic reactors. Entranced by the ease with which images jumped around the videoscreen, Mittelman tried substituting light rays for the program's radiation rays. The result was an effect somewhat akin to latter-day computer graphics.

In the beginning, MAGI subsisted on contract research for government agencies, the Pentagon, and some industrial companies. The jobs generally required Mittelman and his small staff to solve clients' engineering problems using computer technology. But in those insecure years before computer-generated imagery gained the credence it holds today, Mittelman felt compelled to diversify into several areas that still constitute the major source of MAGI's income. For clients that draw their sustenance from direct mail, the firm undertakes the low-tech chore of "merging and purging" their mailing lists. MAGI also developed and sells a computer terminal to help art directors compose those multicolored graphs and charts destined to become the 35mm slides used by corporate managers to illustrate their presentations.

MAGI's contribution to the computer-animation field is a startling process called SynthaVision. This is a method whereby MAGI's animators use solid geometric shapes (e.g., spheres, cones, cubes, cylinders), permanently retained on the firm's proprietary software as the building blocks out of which they construct more complex, three-dimensional objects on a videoscreen.

Animators might want to sculpt an assortment of teapots, let's say. First, they'd graft the appropriate shapes onto each other until their handiwork attained the unmistakable look of teapots. Next, the animators would add color, some texture, and reflectivity. The last is accomplished by indicating the hypothetical source and intensity of light. The computer interpolates how it strikes the teapots' surfaces.

The animators' final task would be to choreograph the teapots' movement. With the mere tickling of a few computer keys, the animators can make the teapots turn, dance around each other, shake, fly apart, or do whatever else their imaginations dictate. Because the computer is monitoring their every move, the teapots maintain the proper shading and perspective at all times as they whirl around the screen defying gravity and every other law of physics.

SynthaVision excels at movement. Its simulated images, on the other hand, meet with a complaint occasionally leveled against the whole industry: Its objects are too perfect; they look synthetic because they don't have enough intricate texture.

It's true that a SynthaVision landscape is generally one whose topography appears engineered down to the last pixel (a square dot of light) on the computer screen. But this evocation of a high-tech cosmos is hardly surprising, since the system evolved out of work done for one of MAGI's military clients. It was never intended to be used solely for artistic purposes. In fact, SynthaVision to this day doubles as a computer-aided design tool, assuming the role of an automated draftsman for manufacturers seeking to significantly reduce the labor involved in creating and modifying intricate technical drawings.

For twenty years, a coterie of nonprofit research institutions did most of the groundbreaking work in the field of computer graphics. To this day, the University of Utah, New York Institute of Technology, Ohio State, Cornell, the University of Rochester, Brown, Rensselaer, MIT, Stanford, and the twin Pasadena institutions—the California Institute of Technology and the Jet Propulsion Laboratory—breed and employ the best and the brightest technoartists in the field. But these bastions of digital wizardry are having more and more trouble retaining their most gifted progeny. The film, television, advertising, and video game industries all crave the stylized symbols and special effects made possible by computers.

The limited but growing demand for *avant-garde* computer graphics has made entrepreneurs out of several of the field's finest magicians despite the fact that it costs a minimum of $2 million to equip a full-scale production studio.

Robert Abel was one of the first to strike out on his own, in 1971. Over the intervening years, his Robert Abel & Associates has won Clio awards for its vibrant and surrealistic TV ads and developed a number of important techniques involving the computer control of cameras during the film process.

Charlie Vaughn, Gene Nottingham, and Don Hudgins opened their Cinetron Computer Systems in 1972 in the unlikely place of Norcross, Georgia. Besides selling the services of its animators and special-effects technicians, Cinetron also markets more than ten machines that automate different stages of the photographic process.

Walt Disney Productions owns eight of them. The equity in this privately held concern is split equally among the three founders and Hal Pearson, a specialist who cast his lot with Cinetron a couple of years after its founding.

Syracuse University lost some of its most promising computer graphics talent when Digital Effects, Inc., was formed in Manhattan in 1978. Seven men put up the initial capital of $1,000; six of them remain this firm's sole stockholders. The president and founder with the most equity is Judson Rosebush, whose muttonchop sideburns make him look like he just stepped out of a Charles Dickens novel. Because Digital Effects has doubled its revenues every year, it's been able to finance its soaring growth out of retained earnings.

The keystone of John Whitney, Jr., and Gary Demos's Los Angeles-based operation, Digital Productions, is a multimillion-dollar Cray scientific computer that they believe will one day "digitize" people, creating fakes so lifelike that a viewer won't even question their humanity. But for the time being, their supercomputer is engaged in the more urgent task of simulating such realistic action scenes that the movie industry can at last dispense with the costly miniature models it's traditionally used in special-effects sequences.

By right of birth as well as achievement, John Whitney, Jr., is joining that closed circle of alchemists long responsible for Hollywood's technical effects. His father, John Whitney, Sr., is credited by many as being the first experimental filmmaker, back in the 1940s, to employ a mechanical analogue computer to manipulate artwork. Whitney, Sr., is also a pioneer in the development of slit-scan photography, which was used to most telling effect to create the time warp sequence at the end of Stanley Kubrick's *2001—A Space Odyssey.* The effect was one of an infinite corridor of light and shapes moving toward the camera at enormous speeds.

Gary Demos was introduced to John Whitney, Jr., through John Whitney, Sr., one of Demos's professors at Cal Tech.

The duo teamed up in 1973, but their first business deal, with Evans & Sutherland, aborted almost before it began. So for the next eight years, they labored as salaried employees in the corporate vineyard of Information International, Inc. There they formed a special division to provide electronic effects to the motion-picture industry.

Donna Kuyper

John Whitney, Jr. (left), and Gary Demos are sitting in front of the Cray-1 supercomputer that is the cornerstone of their Hollywood computer-graphics business. The Cray is one million times more powerful than an Apple II microcomputer and has sixty-seven miles of wire interconnecting its circuits. With the Cray, they plan to simulate scenery, action, and people looking so real that movie audiences won't suspect a computer generated them.

From the start, it was a productive partnership. Whitney and Demos's digital handiwork is well known within the industry. Among their film credits are sequences in *Westworld, Futureworld, Looker, TRON*, and *The Last Starfighter*.

To their peers, the names "Whitney" and "Demos" immediately conjure up the image of Adam Powers, a computer-generated humanoid who does the impossible in a now-famous experimental film. He starts out juggling and ends up removing his head and turning himself inside out. But Adam Powers is by no means a perfect rendering. In fact, he looks something like a smooth plastic doll. His hair is his most obvious flaw. Human hair is the one bodily feature that still has computer animators stumped.

With financial guarantees from Ramtek Corporation, Whitney and Demos tried again and founded Digital Productions in 1981. Its beat is the entertainment industry, which pits it squarely against Lucasfilm's formidable special-effects subsidiary, Industrial Light and Magic, not to mention Lucasfilm's computer development division.

George Lucas has staffed his computer development division with some of the finest craftsmen in the tiny digital-animation field and engaged them in extensive and expensive R&D. The division's stated goal is to create hardware and software products that will automate as many aspects of the moviemaking process as possible—ranging from the purely visual to the audio end of the business. And what's more, Lucas claims he will sell the division's results, since it is his dream to infuse new life into an antiquated industry threatened from all sides by exciting new forms of recreation.

At the same time, Lucas admits that these competing entertainment technologies—such as interactive videodisks, interactive cable TV transmission, and computer animation—hold a fascination for him that he is just now beginning to explore and that he expects his own company to exploit.

PART FIVE

The Information-Age Office

8 Crunching Numbers and Processing Words: *The Work-Station Visionaries*

There are two schools of thought on the office of the future. One, propounded by futurist Alvin Toffler, holds that by the year 2000, most of us will still be working for corporations, but we'll be performing our jobs in our homes. Why buck noisy traffic and waste precious gasoline commuting to and from our employer's office? Why, indeed, since it will be faster and more efficient to execute our white-collar, high-thought-content tasks using a personal computer to "telecommunicate" with the office.

Another futurist, John Naisbitt, says the "electronic cottage" will never become the predominant mode, human nature being what it is. The more society surrounds its workers with high-technology gadgets, the more workers will rebel and choose a "high touch" environment. A high-touch environment is one replete with other people—in short, the good, old-fashioned office.

But given the invention of the microcomputer, that good, old-fashioned office will never be the same again no matter where it's located. In the not-too-distant future, managers and clerical workers alike will be sitting in front of "work stations" dutifully carrying out various information-age chores. These work stations will be souped-up microcomputers able to do everything from processing endless chains of words to showcasing numbers in the form of colorful pie charts and graphs.

And more: A bevy of these work stations will be communicating with each other, linked together in a "local area network." These information webs will encompass work stations within a limited geographic range—a building or cluster of buildings. Through them, the wonders of the future will unfold.

For example, instead of having a mail clerk shuttling back and forth with interoffice memos falling out of his arms, the senders of those memos will be taking advantage of an electronic mail system. They'll tap their message into their work station, press a few buttons, and presto! Over the wires it goes into the electronic mailbox of the recipient.

And someday, so we're told, there will even be work stations that understand our spoken commands, talk back to us, and have a voice-mail capability.

According to the management consultants touting automation's benefits, networks of multifunctional work stations, despite their price tag of over $5,000 for each terminal, still promise to save employers money in the long run. For one thing, expensive peripheral equipment and central data bases can be shared. For another, "knowledge workers" will increase their productivity. A Booz•Allen & Hamilton study revealed that without a work station as helpmate an average of 15 percent of a manager's time is frittered away on nonessential activities.

Finally, automation will do to the office what it's already doing to the factory—reduce the number of employees, especially redundant middle managers.

Needless to say, there is no work station on the market today that offers all the data, text, voice, and image functions just mentioned. Such a wondrous digital workhorse—priced low enough for companies to buy in bulk—is still a figment of our hardware engineers' collective imagination. But they're working on it. Currently we have terminals that can be linked up, but they possess only some of those marvelous abilities. What's more, most of them are still being used today as stand-alone personal computers. But third-generation micros such as Apple Computer's Lisa and Xerox's Star are noteworthy moves in the right direction.

Lisa, the product of an entrepreneurial company, and Star, that of a corporate Goliath, illustrate the forces converging on the "integrated office."

In the one corner scrambling for a tiny place in the sun are hordes of new companies. Most only recently made it out of the garage thanks to generous handouts from venture capitalists. And, with a few exceptions, most will be on their way back to a garage within a few short years, venture backing and the explosive demand for office systems notwithstanding.

The reason is simple: The formidable competition in the other corner. Not only are the newcomers squaring off against two of the country's largest corporations—AT&T and IBM—but also they face a battle from every other office equipment maker and data processing firm on or near the *Fortune* 500 list. And, lest we forget, the predatory Japanese.

Egil Juliussen, chairman of the well-known market research firm Future Computing, estimated that there were about seventy or eighty companies in 1983 jockeying for a piece of the business systems market. He opined that by 1986, the market's structure will be set for the next ten years.

Although the situation may seem hopeless for the entrepreneurial aspirants, a few have already made a dent in the face of the giants. The most successful tactic seems to be to supply the giants through "original equipment manufacturing" (OEM) agreements, as Intelligent Systems Corporation, TeleVideo Systems, Altos Computer Systems, and Convergent Technologies are doing. Such agreements are the reason you've probably never heard of these companies' products. Their electronic handiwork is sold under the names of such well-known companies as Burroughs, NCR, Control Data, AT&T, and Atari.

Intelligent Systems of Norcross, Georgia, and TeleVideo Systems of Sunnyvale, California, have several things in common: Both companies began by manufacturing "smart" (i.e., with built-in memory) color video terminals. They both diversified into the production of total work stations. And both were founded by individuals who put their own hard-earned cash on the line and faced tremendous risks before turning a profit. Actually, neither company founder had any alternative.

Intelligent was founded in 1973 and TeleVideo in 1976, long before the "vulture capitalists" began swarming. But, as it turned out, both companies' self- and debt-financing schemes were disguised blessings. Now Intelligent and TeleVideo's founders are sitting on top of a much larger share of equity than any of their latter-day competitors who formed venture-capital-backed companies. By and large, their entrepreneurial competitors are firms that were founded since 1978 by groups of people rather than individuals. It's the rare member of a founding group who holds more than 5 percent of his or her company's stock.

Charles A. Muench, Jr., the inventor who heads Intelligent, took his company public in 1980 and still managed to retain 30 percent of the stock. TeleVideo's Kyupin Philip Hwang, a Korean immigrant, did even better. When his company went public during the booming new-issues market in the spring of 1983, Hwang's 70 percent equity was worth over $500 million.

Hwang has a refreshingly matter-of-fact attitude about his newfound wealth.

"If I lose everything," he maintains, "I've only lost fifty dollars. That's what I had in my pocket when I arrived in the United States."

Hwang came to the States on a student visa in 1964. In rapid succession, he earned a B.S. from Utah State, an M.S. from Wayne State (both degrees in electrical engineering), and became a U.S. citizen. His business education, in contrast, was strictly seat-of-the-pants, interspersed with an occasional night-school course.

To learn the business basics firsthand, Hwang and his Korean-born wife, Gemma, bought a 7-Eleven franchise in the heart of Silicon Valley and ran it for a year. But all the while, Hwang's mind was toying with product ideas. One finally clicked: What the world needed was a high-resolution color monitor to use with video games. His best idea, though, concerned production. He decided to subcontract out the assembly work to a Korean firm because it would have access to cheap labor.

Despite an initial purchase order from Atari for six thousand monitors, Hwang still had to put up practically everything he owned as collateral for a $25,000 line of credit at the bank. Hwang designed the terminal and worked closely with his subcontractors' employees, sometimes traveling to and from Seoul twice a week. His close attention to detail paid off. By 1980, TeleVideo had moved into first place among the independent suppliers of smart terminals.

Even Charles I. "Chuck" Peddle, the enormously talented engineer-turned-entrepreneur, concedes he could never match Hwang in terms of net worth given his meager 10 percent equity in Victor

Technologies, the company he founded in 1980 with heavy backing from Kidde, Inc., the conglomerate that is its principal stockholder. It's particularly unlikely now, since his third attempt to found a computer company has, like the others, gone awry. In February 1984, four divisions of Victor Technologies—but not its healthy foreign subsidiaries—filed Chapter 11 bankruptcy proceedings.

The financial disarray at Victor isn't too surprising in light of Peddle's track record. His two previous entrepreneurial attempts a decade earlier were short-lived. He says both were a case of being in the right place technologically too early for the marketplace.

Peddle hasn't put forth this or any excuse in the case of Victor Technologies, although observers say he made the mistake of believing his own PR. He squandered millions to staff Victor in the false hope of becoming the No. 2 microcomputer manufacturer in the United States after IBM. In this country, Victor never got anywhere close to second place; but Victor's European operations are a different story. There Victor rose to No. 1 producer of microcomputers, primarily because the company was the first to market a model that processed data in sixteen-bit chunks.

Chuck Peddle also labored under the delusion that his micro's technical superiority *ipso facto* destined it for instant success. Unfortunately, the American consumer frequently places more weight on a manufacturer's reputation, longevity, and customer support than on its products' state-of-the-art performance.

Peddle was right about one thing, though. Victor Technologies did turn out a superb microcomputer. It should have, for electronic design has always been Chuck Peddle's strong suit.

While Jobs and Wozniak are usually credited with making the first fully assembled microcomputer, to the *cognoscenti* Peddle remains the father of the micro revolution. As a MOS Technology engineer in the mid-1970s, he was the chief architect for the 6502 microprocessor that "the boys," fifteen years his junior, incorporated into their Apple II. But "the boys" and their imitators never would have been successful producers of low-cost microcomputers at all if Chuck Peddle hadn't figured out a way to sell the 6502 for $25 when all the other chip houses set $200 as the rock-bottom price.

When MOS was acquired by Commodore, Peddle developed Commodore's personal computer. Commodore's PET and the Apple

II were unveiled at the same West Coast computer fair, but "the boys" shipped theirs first.

Engineering knowledge certainly had nothing to do with Lore (pronounced "Laurie") Harp and Carole Ely's decision in 1976 to found Vector Graphic, one of the older microcomputer firms in the industry. Boredom with their upper-middle-class housewives' existence was more to the point.

Vector Graphic started as a part-time venture, since both women were married with two young children apiece. Their initial product was an 8K circuit board designed by Lore's husband, the engineer-physicist Dr. Robert Harp. They sold them on a COD basis to hobbyists through mail-order ads. The response was overwhelming.

Vector Graphic

Carole Ely (left) and Lore Harp built a thriving microcomputer company around a circuit board designed by Dr. Robert Harp, Lore's ex-husband. With competition from IBM, Vector Graphic is struggling now to regain its supremacy and both women have resigned from the company.

Harp and Ely were delighted and so intrigued with their newfound roles as business tycoons that two years later, their company, based in Thousand Oaks, California, swung into full production with complete microcomputer systems targeted at small-business professionals.

The company's strength is marketing. Because of their nontechnical backgrounds, both women saw their products from a layperson's point of view. As a consequence, they devoted an inordinate amount of their time communicating with dealers and holding the hands of their customers.

Given the steamroller competition from IBM since the 1981 introduction of its PC, the loyalty of dealers remained one of Vector Graphic's few bankable assets by 1982. That was the year the company's growth started to slow markedly. Added to its competitive pressures was turmoil in the executive suite. The Harps' marriage dissolved, Bob Harp left to found his own company, and Carole Ely also divorced, and remarried a company officer. In the fall of 1983, Carole Ely severed all ties with Vector by resigning from its board. Lore Harp followed suit the next spring.

At least for Lore Harp, her personal worth has grown, rather than shrunk, recently. Within months after her divorce from Bob Harp became final, this extroverted, German-born woman made an important decision of her own. In July 1982, she married Patrick McGovern, a man who is worth over $100 million in his own right as the founder and principal stockholder of the privately held International Data Corporation. His computer research/publishing firm is three thousand miles across the country, in Massachusetts. Until her resignation, the couple saw each other only intermittently, at computer shows and conferences.

But could you really expect a high-wattage duo like Harp and McGovern to settle into matrimonial bliss, forgoing all future opportunities to make another fortune? Of course not. In 1984, Lore Harp emerged as the CEO of Pacific Technology Venture Fund, based in San Francisco, a venture-capital firm that will invest half its capital in Japanese entrepreneurial companies. Her husband is a co-investor in the fund.

Who is the youngest person to found a microcomputer company still flourishing after ten hard years' growth? That distinction goes to William M. Wells III. With an engineering degree and four years with IBM in R&D under his belt, he founded Intertec Data Systems in the unlikely place of Columbia, South Carolina, in 1973. He was twenty-four years old. Friends, relatives, and bank loans were his primary sources of capital, leaving him with equity valued at $57 million when Intertec eventually went public.

A Picture Is Worth . . .

Forecasters say the full automation of the office will be a phemonenon of the 1990s. By then, the older generation of computer-haters will be retired or close to it, and the productivity gains needed to justify the expense of such systems will be more apparent.

But few of the psychological and technological obstacles holding back the information-age office exist in present-day offices peopled by the architects of the future.

Architects of the future?

They're our mechanical, electronic, and industrial engineers and other scientifically oriented technocrats. For them, sophisticated graphics work stations offering "*c*omputer-*a*ided *d*esign" (CAD) assistance are already considered *de rigueur* if they are to accomplish their mission—to design outstanding products and physical structures, including huge buildings, in the most cost-efficient manner possible.

CAD systems are to America's approximately one million engineers what word processors are to typists: indispensable.

It's a fact that the mind absorbs the information content of a diagram or drawing up to fifty thousand times faster than it does an array of numbers or words describing the same image. In other words, computer graphics seem to work the way the brain works, especially a scientifically oriented brain. This is why one computer graphics model on a videoscreen is worth a thousand printouts or conventional blueprints in the time, effort, and dollars expended.

Martin Allen, a mechanical engineer by training, was one of the early visionaries to see the awesome potential for CAD equipment. He says the flash of inspiration that culminated in the founding of Computervision, now the leading firm in the industry ahead of IBM and General Electric, came to him during a visit to the Smithsonian Institution in Washington in 1964.

Mesmerized by an exhibit showing George Washington as a land surveyor, it struck him that the tools of Washington's eighteenth-century trade hadn't changed much in two hundred years. Nor had the engineer's, for that matter.

So, Allen reasoned, why not use the emerging minicomputer technology to automate the drafting board, straight edge, slide rule, and compass—the mainstays of the contemporary design engineer's trade? Why not let a computer do the requisite mathematical computations and convert them into a visual image on a video monitor, an

Computervision

On a trip to the Smithsonian Institution, Martin Allen got the idea for computerizing the creative efforts of design engineers. The result was Computervision, an industry leader in the production of computer-aided-design equipment.

image that an engineer could then modify *ad infinitum* and move around the screen for viewing from all angles?

It took Allen four more years to find a partner and get his concept up and running. This is not to say Allen invented CAD. Actually, mainframe computers with graphics capability were available from IBM in the 1950s, but their cost made them prohibitive. No, Allen is something else. Martin Allen is the man who for the past decade has trumpeted the virtues of CAD technology with all the fervor of a missionary. He believes CAD in tandem with something called "computer-aided manufacturing" (CAM) and robotics are the answer to America's industrial productivity crisis. They alone hold the promise of reviving the stagnating industries of the midwestern rust bowl.

"CAD/CAM is becoming as fundamental to industrial manufacturing as the automated processing of financial data has become to business over the last twenty years," he says. "Today, even relatively small companies realize they can no longer perform their normal accounting functions manually. In the same way, manufacturers are beginning to discover they can no longer develop competitive products using traditional drafting boards and T-squares."

Computervision designs both the hardware and the software for CAD/CAM applications. In contrast, most of its competitors concentrate on software development, buy the necessary hardware from the established minicomputer manufacturers, and sell the total package as a "turnkey system."

A typical CAD system consists of a number of engineering work stations linked to a common data base so that updated information can reach all the members of an engineering design team simultaneously. The beauty of the system is that engineers and draftsmen not only design a product on a videoscreen but test it there as well. Instead of building a costly prototype, engineers let the computer analyze how a new automobile engine, let's say, will react when subjected to electronically simulated temperature changes, mechanical stresses, and other conditions that might prevail in real life.

Such digital dynamism doesn't come cheap. A complete system is likely to cost $100,000. But relatively inexpensive alternatives are just now coming on the market from the battery of start-up CAD companies formed since 1980, when the demand for these systems took off.

Entrepreneurial companies in the CAD industry stand a somewhat better chance of surviving than their counterparts making general business-oriented work stations.

First of all, CAD equipment will always be the preserve of "niche" companies because of the narrowly defined nature of the applications. The data base necessary to design automobiles cannot be used interchangeably by architects designing buildings, for example. Second, technically oriented managers are more discriminating buyers. Their engineering background makes them more willing to judge a work station on its own merits. They are less likely to rely on the established reputation of an IBM or be swayed by a snappy multimillion-dollar advertising campaign.

Roughly speaking, the CAD industry is divided into four segments: mechanical engineering applications (e.g., the design of automobiles and airplanes), electrical engineering applications (e.g., the design of integrated circuits), architectural and construction applications (e.g., the design of buildings and other huge structures), and mapping and earth-sciences applications (e.g., site selection and the translation of geographic, environmental, or demographic data into map form).

CAD work stations are welcome productivity tools in all four areas. They're an essential tool, however, for any electronics engineer who designs silicon chips. This application is often referred to as "computer-aided engineering," or CAE.

Because of their complexity, the design of "very-large-scale" integrated circuits (VLSI circuits) is virtually impossible without automated aid. VLSI circuits represent the next generation of chips.

By one definition, a VLSI chip is any IC that takes a single engineer, using manual methods, more than one lifetime to complete. For example, suppose a company, using the traditional methods, wanted to design a computer comprising five thousand integrated circuits. It probably would take up to three hundred man-years over a period of four to six years. But recently it took the Lawrence Livermore Laboratory in California only twenty-four man-months of engineering effort to design such a computer using a new CAE system from a fledgling company called Valid Logic Systems.

Since the design of VLSI circuits and advanced computers is so labor-intensive, it's not surprising that most of the CAD start-up companies of the 1980s are devoted to electronic engineering applications. Among them are Daisy Systems, Valid Logic Systems, Mentor

Graphics, Cadtec Corporation, and Via Systems. Nor is it surprising that most of them are located in the heart of California's Silicon Valley.

In general, these entrepreneurial companies were founded by a group of enterprising engineers with wads of cash stamped "venture capital" sustaining their company's bank account and a relatively meager number of stock certificates in the founders' individual possession.

One young CAD minicomputer maker on the East Coast was also founded by a team of people with venture-capital support. But the broader application for this company's line of hardware products renders the six founders' small amount of individual equity potentially more valuable.

For several months, Apollo Computer was nothing more than an idea kicking around the office of a Palo Alto venture-capital firm. That was until the winter of 1979, when a Sutter Hill Ventures' partner learned that Dr. J. William Poduska was, at the relatively young age of forty-two, casting about for a new project to occupy his MIT-trained mind.

A decade earlier, Bill Poduska had helped found Prime Computer, Inc., a spectacularly successful Massachusetts company that pioneered the concept of high-performance minicomputers ("superminis"). Poduska had stayed on for a number of years to direct Prime's R&D efforts. In January 1980, Poduska was on the verge of a skiing trip when he got the call from Sutter Hill enticing him to start his own company.

On the cross-country flight to meet his backers, Poduska sketched out a business plan. By that afternoon the parties had a deal sealed by a handshake. Sutter Hill agreed to raise the millions needed to start the company. And Poduska flew back to New England to assemble a management team.

Bill Poduska ended up recruiting some of the best talent Prime, Digital Equipment, and Data General—all nearby minicomputer makers—had to offer. The result is Apollo Computer, Inc., a fast-growing company that makes superminis whose power is enhanced by joining them in a network. They're used by scientists doing complex calculations and engineers designing products. One of their chief selling points is that a user can change a design on the Apollo mini's screen in one-thirtieth second. Competing systems can take up to twenty-five seconds to do that.

As for Poduska, he couldn't be happier. He claims starting a company "is the most fun I've ever had over an extended period. It's like being on a rocket with the fuse lit."

9 From Gutenberg to Galactica: *The Space-Age Communicators*

A pundit once remarked that a book is a fifteenth-century invention delivered by nineteenth-century transportation. You wouldn't have to stretch that analogy very much to apply it to other forms of contemporary printed matter.

The typical business letter or report, for instance, is still printed by a nineteenth-century invention, the typewriter. The medium of its distribution isn't exactly modern, either. It's distributed by some variant of the centuries-old courier system or by an updated version of an early-twentieth-century teletypewriter.

Today, though, for the first time in several hundred years, all that is changing. It's changing to the point where the most commonplace methods of transmitting the written word in the future will resemble nothing we've known previously.

By the year 2000, organizations ranging from major corporations to the United States Postal System and its competitors will be utilizing *fully electronic modes of transmission.* They'll be transporting digitized information through wires or bouncing them off orbiting satellites. The information will arrive at its destination faster and cost the same as or less than what we currently pay for today's outdated service.

Likewise, the packaging of information will represent a radical departure from the past. Sure, a lot of our knowledge still will be circulated by means of printed books, magazines, and newspapers. But a lot more will come to us via *electronic data bases* that we gain access to through microcomputers. Data bases, you could say, are the twenty-first century's idea of a library.

The only constant in this whole equation is the written word. To be well informed, people still will have to know how to read. But it's probable that people will read more words from video display screens than from pieces of paper.

The gargantuan molting of AT&T in the early 1980s sent tremors throughout the communications world. Technological forecasters tell us these tremors won't escalate into a full-scale eruption until the last decade of this century. But when they do, we'll witness an industrywide free-for-all of volcanic proportions.

This war of the communications world promises to feature as the principal combatants AT&T and IBM in addition to hundreds of equally determined corporate warriors hailing from the other developed nations of the world. Their battlefield is the planet Earth, although the decisive contests will occur in North America. The prize for the victors: a predominant share of the market for the inter- and intraoffice transmission of voice, data, text, and video images.

The competition between the two behemoths has been festering since 1974. That was the year when IBM entered the satellite business through its partial ownership of Satellite Business Systems (SBS), a joint venture with Aetna Life and Casualty Company. For a steep fee, SBS offers high-speed data-transmission services to organizations that need to send huge amounts of information over long distances. When speed was not a top priority, though, organizations usually used AT&T's land lines.

Now this scenario is changing rapidly.

During the 1980s a new communications infrastructure is being put into place. It affords business customers a wider range of information-delivery options. As a consequence, the older, labor-intensive delivery methods will continue to escalate in price as the newer ones decline in price.

Two factors will drive down the costs of these newer data-transmission methods: The first is advancing technology. The second is the intense competition that will surely develop among the growing legion of companies focusing on this marketplace. Some have products for the futuristic office, while others merely aspire to transport the data in common-carrier fashion.

The companies vying for a sliver of tomorrow's communications marketplace represent five distinct industries: (1) the office-equipment industry—led by IBM, Xerox, Hewlett-Packard, Wang, and

DEC—that plans to sell multifunctional work stations with networking capabilities; (2) the telecommunications industry—headed by AT&T, GTE, MCI, Hughes Aerospace, RCA, Western Union, Graphic Scanning, ROLM, and a host of entrepreneurial PBX makers—with its voice/data switching equipment; (3) several large data-processing and software vendors, including Tymshare and Cullinet, offering products that make it possible for different manufacturers' equipment to exchange information; and (4) the cable TV industry, spearheaded by those companies with urban franchises that are ready to snake their cables through and between office buildings.

The fifth industry is comprised of a group of nascent high-tech firms with unfamiliar names. Their emerging industry is termed "data communications." Why? Because these firms specialize in the design of all manner of computerized hardware and software products that promise to interconnect every electronic device within an office as well as link up offices spaced thousands of miles apart.

Experts estimate that 70 percent of all business communication takes place within the building where an organization is situated or, in the case of a huge company, among the buildings on its corporate campus. At most, only 25 percent of business communication travels over fifty miles.

Local area networks (LANs) are what the name implies: high-capacity electronic connective tissue that joins together all the computerized gadgets within a limited geographic area so they can communicate with each other in the language of that network.

A LAN uses a coaxial cable or telephone line and sophisticated circuitry. A LAN's language is referred to as its "protocol."

A LAN is the evolutionary successor to the "distributed data processing" (DDP) network of the 1970s. A DDP network gave key corporate executives access to a company's central computer, often a minicomputer, through terminals dispersed throughout the office. But by LAN standards, DDP networks are primitive. First, a DDP's "dumb" terminals could send and receive information but didn't have the brainpower to process and store any of it as today's microcomputers can. Second, DDP terminals allowed users only to exchange information with the host computer, nothing more.

In contrast, the LANs networks of the future carry the concept of decentralized computing power one huge step beyond the notion of a host computer in every building with a dumb terminal in every department. LANs make it feasible to have an intelligent work station at every desk. These fully computerized work stations can communicate directly with each other, allowing you to send keyboarded messages to your co-workers via this electronic network as easily as you now communicate with them using the company's telephone system.

Some experts like to say we've entered the era of "embedded data processing" because a miniature computer is becoming an integral —that is, an *embedded*—component of all our office tools, even the office copier and security alarm system.

As you read this, maybe fifty thousand LANs are operating on the premises of commercial users. But International Data Corporation, a respected market research firm, expects the demand for them to zoom from a $200 million market in 1982 to $1 billion by 1988. This is a conservative estimate, according to IDC's competitors. More optimistic forecasters claim a volume of $10 billion by mid-decade.

All agree, though, that once the business community begins to comprehend the productivity gains possible through LANs, the sky's the limit.

Of the twenty to thirty small data communications companies in the field, most were started since 1980 with venture capital. Most of the entrepreneurs have educational pedigrees the length of your arm. And most are clustered within a one-hundred-mile radius of Stanford University on the West Coast, or MIT on the East Coast.

The fledgling data communications firms just now beginning to produce the necessary hardware and software to make LANs commonplace may be premature. Like Ken Miller of Concord Data Systems, most realize their time has not arrived quite yet. Thus they are selling interim lower-technology products to tide them over. In Miller's case, it's modems. He calls Concord's line of modems its "bread and butter" product.

A modem (short for *mo*dulator/*dem*odulator) is a device that enables computers to send and receive data over a telephone line. It connects the digital world of computer communications to the analog world of voice communications. More specifically, it conditions a digital signal so it can be routed through the analog channel of the telephone line. Modems also provide error-checking, fault diagnostics, and autodialing.

The premier manufacturer of modems for microcomputers is another entrepreneurial company, this one located in Norcross, Georgia. Hayes Microcomputer Products was started in 1978 by Dennis C. Hayes, a data-communications engineer in his late twenties. He set up his first assembly line on his dining-room table and did most of the soldering himself. Five years later, his company has practically cornered the market for high-speed personal computer modems and communications software.

Today, modems still are the single most common piece of equipment used in data communications. But as we approach the 1990s and more and more digital office networks—PBX-based LANs, for example—make their appearance, the need for modems will decline.

At this point, confusion is the biggest obstacle holding back the office of the future. It's having an equally deleterious effect on the eager entrepreneurs taking aim at this marketplace. Indeed, there's plenty to be confused about. All told, there are fifty-odd LANs, some developed by the country's largest corporations, others by the emerging small data-communications firms. The architecture and capabilities of all these networks are different.

So what criteria should a potential customer use to sort through the myriad alternatives?

First, he must decide whether to opt for a *single-* or a *multivendor* network. Wang's Wangnet and Datapoint's ARC are examples of single-vendor LANs because they link only their own office machines. Obviously multivendor networks are more flexible, since they enable users to mix and match several manufacturers' equipment in the same LANs.

Then there's the question of bandwidth. At one end of the spectrum are the *broadband* LANs that offer video capabilities such as teleconferencing in addition to other remarkable futuristic services —voice mail and the facsimile transmission of documents. At the other end of the spectrum are the more elementary *baseband* networks. Some are so primitive, in fact, that data and voice transmission don't even coexist on them.

The answer is that there is no clear-cut answer until the communications industry settles on some technical standards to keep the office of the future from becoming an electronic Tower of Babel.

The adoption of LANs standards is turning out to be the great chicken-and-egg dilemma of this decade. The problem is that those competing for a slice of the data-communications pie don't want to choose a network standard until all the office-equipment makers have designed all the products they'll be stringing together. But the office-equipment manufacturers don't want to design any more automated products until all the standards for LANs are enunciated.

The standards stalemate has ramifications in the commercial real-estate industry as well. The electronic office surely will remain forever a figment of our imagination until some definitive decisions are made about the kind of cables builders should suspend in the walls, floors, and ceilings of newly constructed office towers.

There have been several abortive attempts already to settle on a standard. In 1980, Xerox, Intel, and DEC got together to propose Ethernet, a baseband network whose acceptance to date has been far from universal. In late 1982, thirteen companies announced something called an "802 standard." It, too, received a lukewarm reception. Now everyone is waiting for IBM to weigh in with its proposal for an industrywide architecture in 1987.

To complicate matters further, once these hardware standards are adopted, there's another set of software standards issues to resolve. And until some uniformity emerges from these debates, the entrepreneurs in the field will continue to sit on the sidelines.

Some observers contend there will never be a single standard for either hardware or software. Greed being what it is, competing companies will never cooperate to that degree. David Gold, a computer-industry analyst based in the thick of the battle zone in Cupertino, California, asserts: "Companies that are successful over the long term will sell entire network systems, not the pieces" and this factor will work against the adoption of an industrywide standard.

Clearly, it's too early to determine the odds, let alone solicit bets, on the entrepreneurial winners of the data-communications lottery.

If past performance were the only consideration, though, Ralph Ungermann would be the safest wager. Ungermann is a two-time winner in the start-up-company sweepstakes, having launched two major computer corporations before the age of forty.

A veteran engineer who had worked for Intel, Ungermann ran his own software business for two years. Then in 1974 he teamed up with Federico Faggin to found Zilog, Inc., which makes semiconductors and other components for microcomputer systems.

From the start, Zilog was an innovative firm that required regular infusions of cash from its principal backer—Exxon Enterprises, the venture-capital arm of the oil colossus—to fuel its supersonic growth. Therein lay the rub, at least from Ungermann's point of view.

As Exxon's equity stake approached 80 percent, Ralph Ungermann's and Exxon's long-term plans for Zilog began to diverge. Ungermann wanted Zilog to retain its independent status. Exxon and Faggin wanted to merge it with the oil company's other office products ventures. Becoming just another middle manager of just another *Fortune* 500 company was not Ralph Ungermann's idea of a good time.

Proving once again that equity talks, Exxon won out and Ungermann resigned in a boardroom donnybrook that fed the Silicon Valley gossip mills for several months.

When Ralph Ungermann walked out in 1979, his goal was to start a second company. If he had any nagging doubts, they evaporated immediately in the face of the steady stream of phone calls from venture capitalists inquiring about his plans. He didn't crystallize those plans until many months later, after he'd interviewed the officers of a number of large corporations about what future products they needed.

Ungermann had only two requirements: The product should be part of an emerging industry. But the industry, whatever it was, could not be emerging so slowly that his firm would require huge investments before achieving any return.

Ralph Ungermann's extensive market research indicated the need for a network that would interconnect heretofore incompatible computer devices made by a multitude of vendors. To get started building such a product, Ungermann raided his old firm for a partner. That partner is Dr. Charlie C. Bass, the brilliant software expert who was general manager of systems at Zilog.

Together, their company, Ungermann-Bass, Inc., developed Net/One—available in both baseband and broadband versions. Ungermann describes it as "a phone system for computers."

Bass's expertise and Ungermann's entrepreneurial experience are

Ungermann-Bass

Ralph Ungermann (right) left his first company, Zilog, a semiconductor manufacturer, to found his second, Ungermann-Bass, Inc., a producer of local area networks. His partner this second time around is Dr. Charlie C. Bass (left), a brilliant software expert who had been general manager of systems at Zilog.

proving to be a hard-to-beat combination. Ungermann says that knowing what *not* to do the second time around is almost as important as knowing what to do. What you don't do is sell the company's soul to one investor or the venture-capital subsidiary of a large company.

The two founders provided Ungermann-Bass with its initial seed capital. A line of credit from the Bank of America and money from traditional venture-capital sources followed, leaving Ralph Ungermann with 9.7 percent equity and Charlie Bass with 8.2 percent when the company went public in the spring of 1983.

Dr. Robert M. Metcalfe, the cofounder with Howard Charney of 3Com Corporation, is another standout.

In general, Metcalfe's educational credentials are no more spectacular than those of his competitors: B.S. degrees in electrical engineering and management from MIT, a master's in applied mathematics from Harvard, and a Ph.D. in computer science from MIT. But his on-the-job track record speaks for itself: Fresh out of graduate school, he joined Xerox in 1972 and became a principal architect of Ethernet, sharing in its four patents. He also managed the teams of hardware and software engineers who developed Xerox's Star work station.

What better background could a person have to start a company to sell computerized connection devices for Ethernet?

Obviously, the venture capitalists saw it that way, too, for Metcalfe had no trouble raising the necessary seed capital to get 3Com started in 1979. (3Com is an acronym for "*com*puter *com*munication *com*patability.")

Among 3Com's products is an I.E. controller. The letters "I.E." stand for *I*BM *E*thernet. It's a circuit board that makes it possible for IBM personal computers to "talk" to each other over Ethernet. Metcalfe cast his lot with the IBM PC because he guessed that it would become the most popular microcomputer on the market. He pictures his I.E. controller, like an egret on the back of a cow, just going along for the ride. Next up was a similar integrated circuit for the second most popular micro, the Apple. Its designation—the A.E., of course.

Metcalfe, who still looks boyish despite nearing forty, describes himself as "lower middle class upwardly mobile from blue-collar to white-collar." Raised on Long Island, he is the son of a technician who built gyroscopes for the military. He is also the first member of his family to go to college. He'll probably be the first to become very rich as well, at least if ability and stick-to-it-iveness have any bearing on success. He doesn't hide his goal:

"My total net worth is tied up in shares of 3Com. I'm in this up to my ass. I want to conquer the world!" he says, only half joking.

When 3Com went public in 1984, Metcalfe's 14.6 percent equity was disclosed. He's the largest shareholder. Next in the lineup of shareholders are a handful of well-known venture capitalists.

Electronic mail—the dispatch and receipt of typed messages among work stations—may not seem particularly novel in our age. But if electronic mail doesn't raise your eyebrows, it's a sure bet that voice mail will.

Voice messaging systems (or "voice store-and-forward systems," as they are technically known) link telephones in the same way that electronic mail links desktop computers. Simply put, voice mail is a digital method of storing spoken telephone messages and playing them back at the receiver's convenience.

Sounds just like the function of an everyday telephone answering machine, doesn't it? Not quite. While it's true that these systems duplicate some of the features of their low-technology predecessors, they also perform far more spectacular feats.

For example, suppose someone leaves a message in your voice mailbox. After you play it back, you can immediately deliver a reply, and the system will automatically route your response back to the original caller. You can also instruct the system to forward an identical message to one or more recipients at any time you designate. In this way, an executive could, with one phone call, "broadcast" a message to all his subordinates about an upcoming meeting, let's say.

Such built-in flexibility is possible with a voice-mail system because a computer is directing traffic and you can tell that computer to do anything you want. Gone are the audio cassette tapes that store messages on a telephone answering machine. Instead, the voice-mail computer converts a spoken message into bits and bytes, compresses it for more efficient storage, and saves it on a magnetic computer disk. To play the message back, the computer reverses the process.

The main drawback to voice mail is its cost, which ranges from $100,000 to $500,000 for systems capable of handling from several hundred to a few thousand phones. But the price is coming down as the pioneers in the field gradually improve the technology.

Right now, voice-mail systems are separate electronic boxes that attach to an office switchboard. It's expected, though, that state-of-the-art telecommunications equipment manufacturers eventually will embed the technology in future generations of their PBX systems.

In 1980, VMX, Inc., became the first company to market a voice-mail system. Because of its excellent product, this small enterprising firm, located in Richardson, Texas, is considered the industry leader

The name of Gordon Matthews's company is the same as its product —VMX. It stands for his invention, *V*oice *M*essage *E*xchange. VMX is a voice-mail system that digitizes telephone messages and stores or routes them at the sender's bidding.

despite the emergence of Wang, IBM, AT&T, and ROLM as heavyweight contenders for the honor. Its leadership was assured in 1983 when the company received a U.S. patent so broad it virtually covered "the whole industry," according to one newsletter publisher in the field.

VMX—short for *V*oice *M*essage E*x*change, the company's product—was founded by Gordon Matthews, a tall, rangy Oklahoman who enjoys tooling around the flat Dallas landscape on his Honda motorcycle. Everyone agrees Matthews is a genius as an inventor, a man who possesses more technical savvy in his big toe than the average tinkerer carries around in his head. Everyone also agrees that Matthews has no business sense whatsoever, which he proved by losing control of his two previous entrepreneurial ventures.

Matthews got the idea for the Voice Message Exchange in 1974 when, during a visit to a Johns-Manville plant, he happened to pass a huge garbage bin overflowing with telephone messages. The image wouldn't leave him. He was haunted by the thought that each of those messages had cost Johns-Manville about $8 in time, effort, and lost productivity.

Technologically, the Voice Message Exchange is a hybrid of Gordon Matthews's two earlier offerings, the Teleswitcher and the WATSBox. The Teleswitcher, invented in the late 1960s, was the first system for relaying messages between computer terminals. The WATSBox, circa 1972, helped usher in the age of low-cost, long-distance phone service, much to the chagrin of Ma Bell. The WATSBox automatically shunted calls along the least expensive lines. With a special attachment, it also spewed out itemized bills.

At the insistence of his backers, Matthews surrounded himself with seasoned business professionals this third time around. So far, the VMX team has managed to keep disaster at bay while capturing a growing share—70 percent at this writing—of the voice-mail market.

Other embryonic competitors haven't been as fortunate. In late 1982, Threshold Technology, a maker of voice-recognition equipment for computers, filed for reorganization and protection from its creditors under Chapter 11 of the U.S. Bankruptcy Code. Exxon Enterprises' Delphi Communications Systems and a young firm called Voice & Data Systems both bowed out of the voice-messaging business after several years of trying.

* * *

When computers communicate, their signals travel over one of four —or a combination of the four—transmission media. They are: conventional telephone wire, coaxial cable, fiber-optics cable, and microwave.

Local area networks, for example, use either conventional telephone wire (i.e., twisted-wire pair cable), a slow and imperfect highway for digital transmission; coaxial cable, which has the advantage of accommodating video signals as well; or the revolutionary fiber-optics cable, which carries the greatest number of signals of all types at the fastest rates with the least amount of distortion.

Unfortunately, it will be decades before fiber-optics cable completely replaces old-fashioned telephone wire, which currently snakes its way from floor to floor within buildings, burrows into conduits buried under city streets, and hangs from telephone poles lining suburban and exurban roads.

Until fiber-optics cable becomes more prevalent, the best alternative medium to use for sending large amounts of data over great distances is microwave. The typical route microwave signals used to travel was horizontal, following the curvature of the earth. No longer. Today, microwave signals are just as likely to follow a diagonal pathway, signals beamed up to a satellite and back.

At present, most communications satellites operate at the same frequency as ground-based microwave. Systems using this frequency have several disadvantages: Huge dish antennas are necessary to send and receive the signals. The relatively low-frequency signal is especially sensitive to interference, making it imperative to locate earth stations well outside cities, in uncongested areas.

For these reasons, a typical microwave signal has to travel through a telephone wire or coaxial cable for maybe a hundred miles at the beginning and end of its journey.

This may be changing. Some experts claim the next generation of satellites will use advanced technology that accommodates higher-frequency microwave. As a result, smaller dishes are possible, and they can be peppered around cityscapes without fear of disrupting the signal. They predict the day is not that far off when every downtown office building will have an earth station perched on its roof.

Granted, high-speed, computer-to-computer data transmission via satellite is still rare. But a small company in Mountain View, California, is doing its best to make it a more common occurrence.

* * *

You could say that Vitalink Communications Corporation owes its existence to a pile of manure.

Albert L. Horley, one of the company's four founders, lives in affluent Los Altos Hills, a community of twenty-five hundred homes nestled in the foothills of the Santa Cruz Mountains, forty miles south of San Francisco. Soon after the Horleys moved in, Jeanette Horley asked the curly-haired Texan who lives next door if she could use his horse's dung for garden fertilizer. During the ensuing conversation, she discovered the guy was one of those legendary Silicon Valley entrepreneurs who had started his own computer company. Since her husband, Al, was angling to start a satellite equipment company, Jeanette insisted he go meet the nice neighbor.

The neighbor was Jim Treybig of Tandem Computers. Over a few bottles of beer they talked, and Al Horley showed him the prototype satellite receiver he had sitting in his garage. It was made out of sheet metal and was smaller than the standard earth stations of the time.

Horley explained that he'd built his first cheap ground station in Malaysia in the early 1960s for less than $10,000. He was a Peace Corps volunteer teaching physics, and the dish was a class project. The class had actually used it to communicate with the Gemini astronauts.

Jim Treybig was intrigued and instructed some Tandem engineers to offer an appraisal. They confirmed that this dish might be just the ticket for linking widely dispersed Tandem minicomputers into a national network via satellite.

Treybig introduced Horley and partners to his own venture-capital firm, Kleiner, Perkins, Caufield & Byers. Two years later, in 1982, Vitalink was shipping satellite transmission systems at the rate of $5 million a year. Tandem Computers, Inc., was Vitalink's first customer.

It would be inaccurate to claim that Vitalink's journey skyward has been without turbulence. On the contrary, Vitalink initially misjudged its target market—medium-size companies that want to link their branch offices without spending a mint and that are willing to trade off transmission speed to do so. Horley's mistake was assuming that smaller-volume customers would have enough technical savvy to know how to piece together a data-transmission system, buying an earth station here and negotiating for the satellite time and any terrestrial links somewhere else.

What such customers want, Vitalink now supplies: end-to-end installation and transmission service from a single vendor. Such service is possible because of various joint ventures and other intricate arrangements Vitalink has forged with communications giants like Western Union. Vitalink, for instance, swapped 25 percent of its equity for an $11.5 million satellite channel transponder from Western Union.

PART SIX

The Factory of the Future

10 Blue-Collar Automation: ***The Roboticians***

If robots ever start singing, their tune probably will be "Anything humans can do, we can do better. . . ."

Already some of today's computerized machines can, to a rudimentary degree, carry a tune, speak, and talk back, given recent advances in speech synthesis and speech recognition. And it's no bunkum either about robots upstaging human beings if brawn rather than brains is the standard. Even primitive contemporary robots can withstand the dirt and grime and tedium of factory work far better than people ever could.

But before robots take over completely, they'll have to do a lot more than just whistle while they work. The day when robots understand the import of the songs they sing is the day when we humans become their playthings. Not to worry, though. Given the rate at which the science of artificial intelligence is progressing, humans will remain in control for many moons to come.

What is a robot?

This rather obvious question is more complicated than you'd think and has international implications.

The Japanese, for example, classify as robots mere mechanical arms—often called pick-and-place robots—that have little or no intelligence. Their performance is limited to simple repetitive chores, lifting objects and placing them elsewhere. In America we'd call them automated machines. But in Japan they label them robots and use their abundant numbers in their factories to prove that their country leads the world in the use of robots for manufacturing purposes.

The Japanese do have more automated machines at their command than anyone else. They may even have more smart robots doing their bidding. But the intelligent robots they do have are mostly those of American design—built in Japan from technology developed by and licensed from American robotmakers.

On a more fanciful level, there are several other things a robot is not. A true robot bears no relation whatsoever to the animated tin can made famous by such children's films as *The Wizard of Oz.* Nor is a robot a walking humanoid with the brains of a scientific supercomputer on the order of C3PO or R2D2 in *Star Wars.*

Indeed, there's hardly anything anthropomorphic about contemporary factory robots aside from the mechanical arms and handlike grippers on some of them. In fact, most modern robots resemble the same old materials handling equipment we've seen in factories for years—with a couple of differences:

In general, robots are more streamlined and sophisticated-looking, and they're often painted in bright colors to keep any unsuspecting mortals from meandering into their sphere of influence with possibly fatal consequences.

Second, a group of hardworking robots is in sync with some overall, computerized production plan. That's why they do not move in steady unison, the characteristic pattern of traditional automation. Rather, in the fully computerized factory of the future, a silent digital command will make one robot abruptly drop its assigned task and go on to another, or make its robotic neighbor speed up or slow down for no apparent reason.

The Robot Institute of America's official definition of a robot is "a *reprogrammable* multifunctional manipulator designed to move material, parts, tools, or specialized devices through variable programmed motions for the performance of a *variety* of tasks." In short, a robot represents the marriage of advanced computer hardware and programming technology with traditional large-scale mechanical automation.

Leave it to the robotics experts at Carnegie-Mellon University, perhaps the leading school in the field, to devise the most evocative definition. They describe a robot analogously:

A robot is a cross between an old-fashioned machine tool and the human laborer who has operated those machine tools since the advent of the industrial revolution. Like a typical machine, a robot can repeat the same task over and over again with great precision. Like a human operator, it can be taught to do new tasks and use accessory tools to extend its physical repertoire.

The Carnegie-Mellon team concludes, however: "As of today [1981], it is fair to characterize a robot as being much more akin to a machine tool than to a human operator."

The average robot still is long on brute force and short on the ability to sense and respond to changes in its immediate environment. There's no robot anywhere that can do what comes naturally to people—at least not yet. Robots cannot think creatively, reason, or react the way humans do through their senses of sight, hearing, smell, touch, and taste.

A few precocious robots now have crude "sensors," the electronic counterparts of living organisms' sensory apparatus. But to date, the only senses any robots possess are elementary versions of eyesight, giving them basic hand-eye coordination, and "touch," affording them the ability to sense and adjust the pressure they exert.

When a young Columbia University graduate named Joseph F. Engelberger (M.S. in electrical engineering, class of '49) first became enamored with the possibility of programmable robots to man modern factories, he knew such machinery had precursors in the past. In fact, the concept dates back as far as eighteenth-century France, when Joseph Marie Jacquard developed a mechanical loom controlled by punched cards. In the 1830s in the United States, Christopher Spencer invented a programmable lathe whose cutting patterns could be altered with interchangeable cam guides fitted to the end of a rotating drum. Such mechanical controls were standard in the machine-tool industry until the 1950s, when men like Joe Engelberger came up with some newer ways of doing things that incorporated the emerging technology of transistorized computers.

For having the Promethean vision to mass-produce the first industrial robots and enthusiastically promote them, Joe Engelberger is now universally revered as "the father of robotics." He's one of the industry's "grand old men," although he's only in his fifties.

Engelberger's apotheosis is well deserved. Maybe it affords him sufficient psychological rewards to compensate for the millions he'll

Unimation

Joseph F. Engelberger poses with one of his company's Unimate robots. Engelberger founded Unimation, America's first robot manufacturing company, in 1962 and thereby became the industry's spokesman. Robots have not made him a multimillionaire, though.

never realize from robotics. Unfortunately for Engelberger, he launched his crusade about twenty years before the managers who run American factories were ready to heed his message. As a consequence, he traded away most of his equity in Unimation, Inc., to keep it solvent for those years before it finally turned a profit in 1975 and became the industry leader it is today. Such is the financial lot of men who champion revolutionary ideas too soon.

Joe Engelberger credits his fascination with science fiction as a college student and a chance cocktail-party introduction in 1956 for motivating him to pursue doggedly what seemed like an impossible dream.

In a series of stories appearing in the 1940s, Isaac Asimov became one of the first sci-fi writers of any note to picture robots as friends rather than foes of humankind. Engelberger read every one of Asimov's robot tales, storing the import of Asimov's "Three Laws of Robotics" somewhere deep in his unconscious.

Ten years later, while Joe Engelberger was running Consolidated Controls, a tiny Connecticut company that made controls and valves for aircraft, he was invited to a party where he met a man named George Devol. Devol is an inventor of the wild-eyed sort whose obsession at the time was the notion of populating America's factories with robots. At last Engelberger had met his alter ego! Devol, the fabulous character whom nobody trusted with money, would design the robots that Engelberger, the solid-citizen entrepreneur, would build and try to peddle. Engelberger's corporate vehicle for fulfilling this ambition was Unimation.

A convoluted chain of events led to Unimation's founding.

In 1958, Engelberger had a falling out with the owners of Consolidated Controls, who offered to sell the company to him. Engelberger got Norman Schafler, the founder and CEO of Condec Corporation nearby, to invest enough money to gain control. With Schafler's financial support and psychological encouragement, Engelberger gradually moved into the business of making robots. Engelberger's groundwork culminated in 1962 when Unimation was spun off as a separate company jointly owned by Condec and Pullman, Inc.

During its formative years, Unimation was known in heavy-industry circles as "that nutty bunch in Danbury pushing robots." It was also a company riddled with ironies. For one thing, the way Unimation built robots made Santa's workshop look like the factory of the future. "Efficiency" and "productivity" were handy words to throw at prospective robot customers during sales presentations, but they didn't necessarily apply back at Unimation's own plant.

As one former Unimation executive put it, "They seemed to have a hard time getting people who could properly manufacture the robots. There was a tendency to bang stuff together to try and get it to fit."

Needless to say, a continuing shortage of quality-control personnel to inspect Unimation's mostly custom-made products was a persistent problem.

But the most serious problem of all was buyer resistance. The crew-cut, bow-tied Mr. Engelberger evolved into the showman he is today for reasons of survival, for Joe Engelberger quickly realized that a change in public perception was absolutely essential if the manufacturing community were ever to accept Unimation's robots as helpmates.

As Unimation's lead salesman, he discovered that to the practical men who ran factories, the word "robot" conjured up all kinds of frightful images. Nonetheless, he plugged away because, as he says now, "I was wrong on the money, wrong on the time, but I *knew* I was right on the concept."

The first tangible sign that anyone was listening came when General Motors ordered a robot to run a die-casting machine. It did its job well enough that General Motors eventually ordered some more to do other things. The same model robot was reprogrammed to operate a punch press and do spot welding, forging, and machine-tool loading. To this day, American automobile manufacturers are the single most enthusiastic users of robots, employing more of them than any other U.S. industry.

But in the early years, Unimation's sales to companies in other industries remained meager. Finally in 1968, in another attempt to beef up Unimation's balance sheet, Engelberger was reduced to handing over its pioneering robot technology to Kawasaki Heavy Industries, a fateful licensing arrangement that helps explain how Japan achieved its current dominance in the field. Indeed, Japan's hegemony is one of the major obstacles hindering latter-day Joe Engelbergers from assembling a personal fortune in robotics. Unfortunately, only a select few of America's emerging small robotics companies will survive, let alone make their founders superrich. While the Japanese now have about 40 viable robot producers out of a total of approximately 130, we have closer to 12 viable firms out of some 50 or 60 *in toto.*

It's true that the Unimation-Kawasaki deal seemed fair enough at the time: Unimation sold Kawasaki the rights to manufacture Unimation robots in Japan, the world's fastest-growing market, in exchange for royalties and the rights to any technology Kawasaki developed. Unimation also gained patent protection, since Kawasaki would see to it that no Japanese competitor made knock-offs.

Fine, except for two sticking points: (1) Kawasaki never developed any technology worth importing and (2) Kawasaki got so good at churning out and selling Unimation-designed robots that today it competes successfully against its licensor in the United States. One example: When Nissan Motor Company outfitted its Datsun truck plant in Smyrna, Tennessee, it bought the majority of its 220 robots in the Land of the Rising Sun, from Kawasaki, which is now one of that country's leading robot producers.

In the 1970s, several things happened that began to make robots look attractive, even to skeptical manufacturing executives. Labor, raw materials, and energy costs were soaring at the same time that the price of robots was declining, albeit slowly, thanks to advances in computer technology. The same advances were also shrinking the size of robotic components and increasing their reliability.

To the disgust of many manufacturers, the 1970s was also the era of OSHA (Occupational Safety and Health Administration) and EPA (Environmental Protection Agency). Moreover, the cost of medical disability benefits was increasing exponentially, driving home the point that an injured employee requires more expensive doctoring than a broken robot.

A 1981 survey by the Robot Institute of America bearing the ominous title "The Decline of Productivity and the Resultant Loss of U.S. World Economic and Political Leadership" purported to document the extent to which America's once-mighty manufacturers had fallen behind their global competitors. One indicator cited was a drop in capital investment per worker. In a span of six short years (1967–73), West Germany had increased its capital investment per worker by $395, Japan by $133, while the United States had actually *decreased* its investment per worker by $38.

That study and others like it in the early 1980s drove home the point to American management and workers alike that unless something drastic was done soon, the country's traditional smokestack industries (e.g., steel, automobiles, machine tools, etc.) would eventually wither and die, succumbing completely to foreign competition.

The call to arms went out. Who or what would revitalize those threatened industries?

Robots were seized upon as part of the solution. Granted, they are expensive—most still cost at least $50,000 each—but they toil away at humdrum tasks twenty-four hours a day, don't take coffee breaks, belong to troublesome unions, or file lawsuits alleging sexual harassment or discrimination. Even better, they seem to thrive in environments unfit for human habitation, and they never complain when they have an accident.

The call was answered by two categories of robotmakers: established companies looking to diversify into leading-edge product lines, and infant companies fathered by fledgling entrepreneurs.

Throughout the 1970s, one after another of America's old-line industrial companies, dubious about their own long-term survival in a rapidly changing world, began shifting their focus toward more promising areas. Robotics is one of these areas.

Cincinnati Milacron, with its roots in the nineteenth century when it was known as the Cincinnati Screw & Tap Co., is foremost among them. The management of this family-controlled company has been handed down through three generations of the Geier family. All the while, Milacron was expanding into the largest machine tool manufacturer in the country.

The robots Milacron started developing in 1968 remained awash in red ink until as late as 1981, when the business showed its first slim profit. Over that same period, Milacron moved into the No. 2 spot among all U.S. robotmakers, and management expects robot sales and the design and installation of complete "flexible manufacturing systems" to equal the company's traditional machine tool sales by 1988.

PRAB Robots, Inc., of Kalamazoo, Michigan, dates back to the 1950s, when it was known as PRAB Conveyors, a manufacturer of equipment used to transport and process scrap metal. In 1961 it was acquired by Charles Larson and John Wallace. Larson and Wallace had both worked for a company competing against PRAB, only Wallace had been fired from his top-management job there. Thus revenge was among Wallace's motives for buying PRAB, transforming it into a successful company and diversifying into robot production. Today it's No. 3 in market share in the robotics industry.

Wallace seems to take the most pride in PRAB's diversified list of customers—the "one-here, two-there market," as he puts it. None of the company's mostly small-business customers has more than twenty PRAB robots, nor are many of its clients involved in the automotive industry, which is unusual.

PRAB's line of robots can manipulate payloads ranging from ounces up to one ton. They toil away at such applications as materi-

PRAB Robots of Kalamazoo, Michigan, originally made conveyors and other equipment to handle scrap metal. In 1961 it was purchased by John Wallace (right) and a partner and evolved into a robotmaker. Walter K. Weisel (left), a former executive at Unimation and Cincinnati Milacron, was enticed with a generous block of stock to manage PRAB Robots.

als handling, spot welding, forging, machine loading, and parts transfer.

When PRAB went public in 1981, the world learned that Jack Wallace owned a 37 percent stake in the company, which dropped to 19 percent after the offering. Wallace's right-hand man, Walter K. Weisel, a former employee of both Unimation and Cincinnati Milacron and who is extremely active in the industry trade association, was the next largest stockholder. Weisel owned 20 percent of the shares before the offering and 17 percent afterward.

A Dale, Indiana, company with the unlikely name of Thermwood Corporation made plastic components until 1977, when it sold that business for $5 million and used the proceeds to launch a line of hydraulic industrial robots. Its founder, Ken J. Susnjara (pronounced "Susnura"), has devised a growth strategy he calls "private labeling" that sets Thermwood apart from its competitors. Thermwood sells its robots to original-equipment manufacturers (OEMs), who resell them under their own corporate names.

A Thermwood robot, for example, is sold through a Kansas manufacturer of web-fed printing presses. This robot's assignment is to unload printed materials from the end of the press and tie them together in bundles. Another one does arc welding and is marketed by a supplier of equipment to that industry.

Arc-welding robots are also a specialty of Advanced Robotics Corporation, the Hebron, Ohio, firm that was created in 1979 as part of a leveraged buy-out of Automated Welding, an Air Products & Chemicals division. The sale was engineered by Ronald C. Reeve, Jr., in exchange for 10 percent equity. He's the same man who had headed the division since it first started making robotic welding and inspection equipment for the Apollo space program in 1969.

The jobs robots do are classified by level of complexity. Arc welding, inspection, and assembly work are among the most difficult because they require the highest degree of intelligence.

Arc welding, for instance, requires a good eye and a steady hand. It also requires the stamina of a bull, since it's a hot, dirty, unpleasant job any sane human being would be delighted to forgo. Because the work is so oppressive, most human welders have their torch on only about 30 percent of the time. Moreover, they must wear heavy protective clothing and face masks to fend off the continual shower of hot sparks and choking smoke. A robot who does arc welding, in contrast, keeps its artificial eyes glued to the seam it is welding and its torch going about 90 percent of the time. This is why arc welding is expected to become robotmakers' single biggest market in the near future.

Only recently have robots begun undertaking arc welding and other higher-complexity tasks. This is possible because of the development of "artificial vision" (also called "computer vision," "machine vision," and "intelligent vision"). Artificial vision is the missing link, the key faculty that, along with a tactile sense, will enable robots eventually to mature into second-generation helpmates.

The principal inventions necessary to reproduce human sight have been around for several decades. They are the television camera and the computer, the machine equivalents of the eye and brain, respectively. But mainframe computers were too cumbersome and expensive to incorporate into a vision system. What was needed was a computer small enough, powerful enough, and cheap enough to process vast amounts of visual information swiftly and accurately, even when lighting conditions are poor. Such a digital dynamo—the microcomputer—became a reality only in the late 1970s.

The typical artificial-vision system consists of a TV camera connected to a mighty miniature computer. It is software, however, that differentiates the competing vision systems. The software tells the computer how to analyze and interpret the visual images generated by the TV camera. Thus, a company's vision system is only as good as its programming.

About twenty companies are now producing their own proprietary vision systems. Many of these companies are traditional robot producers. Others have no dual loyalties. Their efforts are 100 percent dedicated to the goal of advancing the technology of simulated sight, a technology that has a range of other applications besides robotics.

By and large, the small start-up companies in the latter category were founded by defectors from university and high-tech companies' research labs. John W. Artley is an exception. He's the former head of his own advertising agency (with degrees in English, commercial art, and journalism) who used the invention of two University of Virginia medical equipment researchers to start his company in 1972. After years of refining its technology and studying the marketplace, Artley's Object Recognition Systems (ORS), headquartered in New York City, now sells a series of systems ranging in price from $12,000 to $80,000.

ORS's most popular application is visual inspection, a menial job

John W. Artley headed his own advertising agency before he founded Object Recognition Systems. His company makes artificial vision systems. Here he demonstrates ORS's Ibot-1 system.

that is performed currently by more than one million American assembly-line workers.

Here is another case where robots do it better. Routine inspection tasks are incredibly tedious and unrewarding—not to mention sleep-inducing. The most conscientious workers find it difficult to concentrate on the details of moving objects for long periods of time. Video sensors can inspect moving objects on a production line much faster than human beings and make quicker and more consistent judgments about which of those finished goods fail to meet minimal quality-control standards.

Machine Intelligence, Inc., of Sunnyvale, California, is another company whose technology had its origins in somebody else's research lab.

Machine Intelligence was founded by Dr. Charles Rosen, who worked in SRI's prestigious Artificial Intelligence Center for almost twenty years developing various robots and sensors for use in mass production. Rosen and colleagues worked for eighteen months in the garage behind his house developing the prototype of their vision system. (In case you haven't noticed, garages are to recent high-tech mythology what castles are to the old Arthurian legends.)

Once the prototype was completed, venture capital wasn't a problem, but a lack of manufacturing customers was. The largest class of purchasers remained R&D firms and R&D units of companies and universities—curiosity-seekers, in other words, rather than real users.

When the right kind of sales failed to materialize, Rosen was relegated to board member and chief scientist, and a team of professional managers, recruited by the company's investors, took over. Soon thereafter, the new management signed an agreement with Yaskawa, a Japanese robot manufacturer, with the object of giving Yaskawa's blind robots the ability to see.

There are two regions of the United States that are hothouses for infant artificial-vision companies, and for robotics companies in general. Interestingly, neither of them is anywhere near Silicon Valley or the Sun Belt paradises. The Detroit area, because it cradles the robot-infested automobile industry, is one. It is the home of Perceptron, one of the handful of viable companies in the embryonic computer-vision industry. Boston and environs is the other. Octek, Inc., and Cognex Corporation hail from there.

John Trombly and Arthur Fox of Octek got the idea for their product when they were working together at Hewlett-Packard's Medical Products Division on an ultrasound imaging system. They decided to adapt a version of that technology for use in industry, and Octek was born in 1978.

Cognex Corporation's optical character recognition system was developed by Dr. Robert J. Shillman, who started working on a precursor of the present system while studying for his doctorate at MIT. Later, as a consultant to the U.S. Department of Defense, he helped create a computer vision system that can scan and read Russian-language documents. So, he figured, why shouldn't the world be blessed with a similar system that can read English? Read it can. The Cognex system probably reads better than a quarter of the graduates of American grammar schools.

But machine vision has a long way to go before it makes anybody superrich. By the end of 1982, the collective sales of the industry were a poverty-level $19 million.

Laura Conigliaro, Prudential-Bache's respected robotics industry analyst, put that figure in perspective. "Nineteen eighty-two was a transition year for machine vision. Only then did the products move out of the laboratory and into the beginning stages of user acceptance."

All this entrepreneurial activity would be perfectly normal if it weren't for the glacial pace at which American manufactuers are investing in the new technology. Robots have always attracted more media hype than American buyers. As I write, there's still a bigger market for robot toys than for the real thing. Consequently, some U.S. inventors and many venture capitalists are beginning to wonder if they've been had.

In the early 1980s, eager investors poured millions into what was touted as America's next high-growth industry. Millions are what it took to launch this capital-intensive industry, since a minimum of $5 million is essential to any start-up company gearing up for robot production.

Indeed, at one point during the robotics frenzy of 1980 and 1981, investor avarice was so uncontrolled that Mitchell Weiss, one of two former Unimation employees who founded United States Robots, Inc., in 1980, joked, "If you go down to Wall Street and yell 'robot,' people will throw money at you so fast you will have to duck." Weiss and his partner tried it and went home with close to $2.5 million in venture capital. Then two years later, they sold out completely to an electrical products company for $8.5 million.

But the circus came to an abrupt halt in 1982 when several of the giants—General Motors, IBM, General Electric, Westinghouse, Bendix—stepped into the robotic arena with the intent of buying up, or otherwise eliminating, as many of the new little-guy companies as possible. The appearance of General Motors on the list was particularly chilling to investors, since GM had been the single largest *consumer,* the world over, of other companies' robots.

Suddenly that euphoric enthusiasm on Wall Street turned into dismay and cynicism. Robotics entrepreneurs who hadn't gotten their money while the getting was good were plumb out of luck, at least for the time being.

As things developed, American entrepreneurs were more ill-fated than they thought. Thanks to the ever-present Japanese, those minnow-sized robotmakers that were so inclined couldn't even count on being swallowed up, for a nice price, by the corporate whales in the industry. Instead, the whales ignored them and signed licensing agreements to market the cheaper robots of the Japanese. IBM signed with Sankyo Seiki, General Electric with Hitachi, Bendix with Yaskawa. General Motors formed a joint venture with Fujitsu Fanuc Ltd. to build robots in the United States sporting a GMFanuc logo.

Westinghouse Electric Company was one of the few to buy American. It purchased Unimation, the industry leader.

CAD + CAM + Robots = CIM

General Electric may not have acquired any American robotics companies, at least not yet, but it's been actively buying up other U.S. companies that represent pieces of the factory-of-the-future concept. GE has made it clear it intends to dominate the industry, becoming a one-stop shopping center for "steel-collar workers" as well as for CAD work stations, local area factory networks, and all the other high-tech paraphernalia necessary to automate a manufacturing facility fully.

What is this thing blithely referred to as "the factory of the future"?

It's more than robots, that's for sure. Robots are, in fact, only one—and by no means the most important—component of tomorrow's factory. Robots can be used in a stand-alone mode to substitute for human labor, the way most are utilized today. But to realize their full potential as programmable machines, robots can also be incorporated into "computer-aided manufacturing" systems, as most will be utilized tomorrow.

The factory of the future sounds too good to be true. In this manufacturing utopia, a series of powerful and intricately programmed computers, linked together in a vast network, direct the manufacture of a product from its inception as a prototype on a design engineer's CAD work-station screen to its completion, perhaps a hundred steps later, as it rolls off a factory assembly line ready for shipping.

Using this system, a design change in the product is accomplished merely by reprogramming the army of computers. No one has to physically add or subtract any machines from the factory's arsenal. Needless to say, such a system would shorten by many months the turnaround time from a product's conception to its production.

As always, there are lots of buzz words floating around to describe this as-yet-unfulfilled ideal: Computer-aided design (CAD) when it's linked to computer-aided manufacturing (CAM) in one continuous-flow system is known as "computer-integrated manufacturing" (CIM), otherwise known as "flexible manufacturing." The end result is the "computer-integrated factory" (CIF).

The gibberish comes to us compliments of the machine-tool industry, all part of a high-pressure attempt to educate the marketplace to accept this horrendously expensive technology. It's a technology that still does not exist, mind you, even in Japan. The Japanese, however, have more of the building blocks of the technology up and working than we do.

The incompatibility of computer hardware remains one of the greatest stumbling blocks holding up the factory of the future. Paul Wright, a professor at Carnegie-Mellon's Robotics Institute, likens the problem to "trying to run a restaurant with a Russian chef, a French baker, a Chinese dishwasher, and an Egyptian waiter, each of whom speaks only his native language and is hard of hearing."

Lately, Phoenix Digital Corporation has been endeavoring to remedy this problem. But that was not its founder's original intention when he hung out his shingle in 1978. Paul F. Rollins just wanted to develop a comfortable engineering consulting services business with major manufacturers as his clients.

Phoenix Digital

When Paul Rollins hung out his company's shingle in 1978, he thought he was starting an engineering consulting business. It turned out to be a manufacturing business. Phoenix Digital now builds computerized fiber optics networks linking all the robots in factories.

One company Rollins hoped would become a client was General Motors, since he'd acted as a consultant to GM in his previous job at Motorola. GM came through with an assignment, but it was more than Rollins had bargained for. The assignment was to build GM's first factory computer network, using fiber-optics cable, in its Oshawa, Ontario, body assembly plant.

Using the new glass-thread technology was the least of Rollins's problems. It was making GM's jumble of computers and robots "talk" to each other intelligibly that posed the challenge. The GM contract instantaneously transformed Phoenix Digital from a mere consulting firm to a manufacturing company in its own right. After acquitting itself admirably on that first formidable job, Phoenix Digital now has contracts from other industry giants for similar plant monitoring systems. In particular, it's building a local area network for that infamous Nissan Motor plant in Smyrna, Tennessee, the one outfitted with all those Japanese knock-offs of Unimation robots.

Control Automation, Inc., was founded in 1980 by two Ph.D.s who worked together at Western Electric in New Jersey. Although the company's initial products are an assembly robot and an artificial-vision system, Drs. Gordon I. Robertson and Anton F. Rodde's long-range goal is to fill some of the yawning gaps that presently exist in CIM technology.

Control Automation's seed capital came from Frederick Adler, the well-known New York City-based venture capitalist who hauled in his first big bundle of cash in the late 1960s as a backer of Data General, the minicomputer dynamo. Adler drove a hard bargain with Robertson and Rodde. In exchange for the financing, Adler pocketed two thirds of Control Automation's equity.

In that same year, three states to the north in Massachusetts, Automatix, Inc., was started by a group of eight entrepreneurs convened by a talented mechanical engineer with one enormously successful company already to his credit.

French-born Philippe Villers has degrees from Harvard and MIT and was a cofounder of Computervision, the industry leader whose work stations fuel the CAD side of the factory of the future. Villers had wanted Computervision to diversify into robot-making. When it didn't, he spun off his own company and took some Computervision employees with him. Computervision responded with the usual lawsuit, featuring the usual allegation—"wrongfully appropriated" proprietary ideas. A jury didn't agree. Villers was exonerated of all charges.

Given the limitations of current technology, Villers and his team have set up a state-of-the-art production plant in Billerica, Massachusetts, that could serve as a model for Automatix's customers. Its centerpiece is the ManMan Information System developed by Sandra Kurtzig's ASK Computer Systems (see Chapter 4). Utilizing ManMan, Automatix has been "lean and mean from the start," according to a company official. "We probably have thirty percent fewer people than electronics companies with paper systems."

Although Villers brought Automatix to life to sell sophisticated robots and a computer vision system, he clearly has his eye trained on the CAM side of the equation. His eventual goal is to market a turnkey CAM system that integrates five key elements: (1) robots; (2) computers for control; (3) advanced sensors for feedback; (4) software providing a high degree of intelligence; and (5) the necessary fixturing and materials-handling equipment.

"In our view," he says, "what's needed is better sensors and software, not better robots. We can license excellent existing robot technology. Instead, we're concentrating our resources on those areas where incremental improvements will bring the greatest benefits."

Such a statement coming from Villers is not surprising. He's a man with the prescience to recognize that certain global socioeconomic trends make the computer-integrated factory inevitable despite the cost and herculean difficulties involved.

John Goodman

Philippe Villers cofounded Computervision with Martin Allen. But robotics, rather than CAD equipment, was Villers's real interest. In 1980 he and eight other entrepreneurs started Automatix. Villers is pictured with one of his company's robots.

Whereas product lifetimes were once measured in years—decades, in the case of Henry Ford's Model T—this is no longer true. Not only do TV-saturated audiences get bored faster with the same old model, but people are also becoming more individualistic in their tastes. In addition, there's the accelerating phenomenon of technical obsolescence, which transforms many products into has-beens before they even hit the retail shelves.

In manufacturing terms, these trends necessitate less mass production and more batch processing. Flexible manufacturing systems facilitate this transition toward shorter production runs because they make it possible to produce smaller and smaller quantities of goods cheaply.

In the first quarter of this century, Henry Ford, Sr., ushered in more than automobiles. He introduced the era of Detroit-style "hard automation," which has been with us ever since. The machinery subsumed under this generic term churns out look-alike objects quickly and efficiently. But try to get all that fixed equipment—conveyor belts, numerically controlled machine tools, and the like—to do anything else for an hour or two and you're in trouble. Such machinery can do only the one job it was designed to do when it was installed. If you want it to do something else, you'll have to get yourself a new machine—or radically alter the one you have.

It is the inflexibility of hard automation that gave rise to this rule of thumb: The shorter the production run, the greater the cost of the finished goods.

The factory of the future, in contrast, is characterized by "soft automation" (i.e., software-driven machines such as robots). It's a flexible form of automation that can easily accommodate the shorter production runs that will eventually become the norm.

Thanks to robots and flexible manufacturing, in the future there will be fewer jobs that require bodily force and more that require cerebral involvement. Naturally, the elimination of as much human musclepower as possible has American workers and their labor unions worried.

Just how many people will be replaced by computerized machinery?

The academics studying the question claim the number of workers won't be as large as people imagine, nor will the evolutionary displacement process be as wrenching.

A Carnegie-Mellon study estimates that robots could replace one million workers by 1990 mostly in the automotive, electrical-equipment, machinery, and fabricated-metals industries. And by the year 2000, another two million jobs could be lost. It concludes:

"By 2010 or so, it is conceivable that more sophisticated robots will replace almost all operative jobs in manufacturing (about 8% of today's [1982] workforce), as well as a number of skilled manufacturing jobs and routine non-manufacturing jobs. . . . Even though several million jobs in the current manufacturing workforce are vulnerable to robotization, the transition seems hardly catastrophic on a national scale, provided new job entrants are properly trained, and directed . . . to learn marketable skills other than welding, machining, and the other operative tasks that are now being robotized. . . . The jobs would not disappear all at once, and robot manufacturing, programming, and maintenance itself will provide some new jobs."

According to Carnegie-Mellon's Professor Robert Ayres, roughly one technician is required to oversee every five steel-collar workers. On the other hand, he's also quick to point out that, over the next decade, most robot technicians will be recycled blue-collar laborers who are still working away, albeit at more responsible jobs, in the same production areas where their robot charges are now toiling.

The workers who have been upstaged by robots so far have mixed feelings about it. Many are as relieved as they are angry.

One man, whose old job consisted of spray painting appliances all day long, says, "At first I was bitter. How would anyone feel when a bucket of bolts comes in and takes their job away? But I didn't like breathing the paint. That robot may be saving my life."

PART SEVEN

Entrepreneurs Who Are Candidates for the Richest Americans List in the Year 2000

There are 375* candidates comprising my list of entrepreneurs who, at this writing, have a running start at amassing an awesome personal fortune by the year 2000.

What will constitute superwealth in the year 2000?

If you assume that an individual with a net personal worth of $50 million or more in the year 1980 was "very rich"—an assumption I made when I compiled the lists of the richest living Americans in *The Very Rich Book*—then that same individual's personal worth would at least have to keep up with the rate of inflation if he or she were to remain in that charmed circle by the year 2000. Of course, the rate of inflation over the next twenty years is anybody's guess. Thus our definition will have to remain fluid.

There's another statistical question germane to this list: What is the upper age limit for entrepreneurial candidates seeking inclusion on a list of *living* people in the year 2000?

The answer lies in the *Life Tables for the United States: 1900–2050,* compiled by the U.S. Department of Health and Human Ser-

TABLE 7 Financial Criteria for "Very Rich" Status in the Year 2000

Average Inflation Rate, 1980–2000	*Minimum Net Personal Worth Needed to Qualify as "Very Rich" in the Year 2000†*
2 %	$ 74.30 million
4 %	$ 109.56 million
6 %	$ 165.51 million
8 %	$ 246.34 million
10 %	$ 366.40 million
12 %	$ 544.63 million
15.1 %	$ 1 billion

†Assuming at least $50 million net personal worth was needed to qualify as "very rich" in 1980.

*When two or more names appear together on the list—for example, in the case of husband-and-wife teams or brothers involved in joint business ventures—they are counted as one individual.

vices (Social Security Administration, Office of the Actuary), published in September 1982. Table 5 in this booklet indicates that a man who is sixty years old in 1982 has a good chance of living for another 17.65 years—in short, to the age of seventy-eight, the age he will be in the year 2000. The prospects for women are even better. They have another 22.56 years to look forward to.

TABLE 8 Industries Represented on the Year 2000 Richest American Entrepreneurs List*

Industry	*Percent of Candidates Whose Wealth to Date Emanates from the Industry†*
computer hardware and components	20%
computer software/data processing	10%
entertainment (games/movies/TV programming/recreation)	8%
telecommunications	7%
real estate	7%
robotics and CAD/CAM	6%
medical/health care	6%
finance (insurance/brokerage/leasing/private investment)	6%
biotechnology	5%
broadcasting/cable TV/direct-broadcast satellite	5%
energy/oil/gas	4%
publishing/market research/public relations/advertising	4%
computer retailing and wholesaling	3%
venture capital	3%
food	2%
clothing	2%
personal-care products and services	2%
education	1%

*The list contains the names of 375 individuals. Names that appear together on the list—such as husband-and-wife teams, or brothers involved in joint business ventures—are counted as one individual.

†When added, the percentages on this table exceed 100 percent because several entrepreneurs have interests spanning more than one industry.

As a consequence of the data in these actuarial tables, my list of candidates does not include any men who will be older than seventy-eight in the year 2000. The female age ceiling is a moot point, since the few women on the list are relatively young.

The following list is not limited solely to candidates in the dozen or so high-growth industries described in depth in this book. I have also included entrepreneurs who are well on their way to amassing fortunes in such perennial industries as real estate, energy/oil/gas, finance, food, and clothing.

A breakdown of the industries featured on the list appears in Table 8.

You might ask why an industry such as biotechnology, which holds so much promise and will grow so swiftly over the next couple of decades, does not have more entrepreneurial candidates representing it on the list. There are several reasons:

The gene-splicing industry is extremely young and heavily populated by university professors and medical doctors who, by and large, have no previous business experience or entrepreneurial track record. Moreover, to date, it is the rare biotechnology firm that has shown a profit. However, because of the industry's enormous potential, a number of *Fortune* 500-size companies hold major chunks of equity in these "entrepreneurial" R&D firms.

But ask yourself: How long will such firms remain "entrepreneurial" should they experience a big success with some long-awaited, genetically engineered product? Not long, I would speculate. At the point of a firm's greatest triumph, its heretofore benevolent corporate backer will suddenly swoop down like a vulture and show its true mettle as the avaricious and acquisitive capitalist entity it is. Its precocious protégé will find itself, almost overnight, simply another division of a huge industrial giant.

Moral: Not all emerging high-growth industries throw off substantial numbers of self-made multimillionaires. It depends on the industry and its historical context.

TABLE 9 Candidates for Richest American Entrepreneurs in the Year 2000

Candidate (Principal Residence)	Age in Year 2000	Source or Sources of Wealth to Date
Frederick R. Adler (New York City, New York)	75	He made his first millions as a founder of Data General Corp. (minicomputers). He used that grubstake to form Adler & Co. (venture capital), which backs other people's entrepreneurial ideas.
John C. Alderman, Jr. (Norcross, Georgia)	65	Alderman's Digital Communications Associates makes local area networking systems.
Joseph L. Allbritton (Houston, Texas)	75	Houston real-estate purchases when he was a young man provided the foundation for his empire. His Perpetual Corp. now encompasses TV stations, newspapers, and financial institutions.
Dick P. Allen (Irvine, California)	55	Allen joined forces with James M. Sweeney in 1979 to start Home Health Care of America, a company that provides advanced medical services and products to the chronically ill in their homes.
Martin A. Allen (Bedford, Massachusetts)	69	He cofounded the company that's risen to a leadership role in the increasingly vital CAD/CAM field. Computervision's other founder, PhilipVillers, left to start a firm in a related field—robotics.
Paul G. Allen (Bellevue, Washington)	47	As a teenager, Allen and a high-school buddy named Bill Gates got their first paid computer assignment. They founded Microsoft while they were still in college. Today their company is vying for first place as the nation's premier systems microsoftware publisher.

Carlton Amdahl (Santa Clara, California)	48	Carl Amdahl may be living proof that the ability to design computers is an inherited trait. While he was still a college student, Carl designed his first mainframe for the start-up Magnuson Computer. In 1980 he teamed up with his father, Gene, one of the best designers in the industry, to start Trilogy Ltd., the manufacturer of an IBM look-alike mainframe. He resigned four years later when the company was in trouble.
Gene M. Amdahl, Ph.D. (Santa Clara, California)	77	Embittered because he lost control of the company that bears his name, Gene Amdahl was determined to have a second go at it. The financing for his No. 2 company, Trilogy Ltd., was arranged to ensure that he and his son would always retain the controlling interest. But his plans went awry when design snafus placed the company's future in doubt.
Martin A. Apple, Ph.D. (Daly City, California)	61	Apple left his first entrepreneurial venture—a biotechnology firm called International Plant Research Institute—to start a firm, EAN-TECH, that uses computer-aided design methods to construct tailor-made molecules.
John William Artley (New York City, New York)	62	His Object Recognition Systems is considered a leading contender in the emerging artificial-vision field.
Thomas H. Aschenbrenner (Allen, Texas)	53	With three partners and heavy backing from Exxon, he cofounded InteCom to produce advanced telephone equipment for combined voice/data transmission.
Mohd A. Aslami, Ph.D. (Sturbridge, Massachusetts)	53	He left Corning Glass to start a fiber optics manufacturing business, SpecTran Corporation.

TABLE 9 *(Continued)*

Candidate (Principal Residence)	Age in Year 2000	Source or Sources of Wealth to Date
Jesse I. Aweida (Louisville, Colorado)	69	Storage Technology, his co-venture with several partners, makes peripheral equipment compatible with IBM mainframe computers. In 1984, his once high-flying company filed for bankruptcy protection.
William T. Baker III (Sausalito, California)	45	Fresh out of Indiana University, he founded Information Unlimited Software (IUS), which sells popular programs for IBM's PC. In 1983 he sold IUS to Computer Associates International for $5 million plus an incentive bonus of another $5 million based on future performance.
Leon Barstow (Tucson, Arizona)	59	This biochemist founded Vega Biotechnologies to develop drugs. When that goal eluded him, he moved into two more promising businesses: the manufacture of computerized "gene machines" and supplying laboratories with biochemicals.
Art Bartlett (Santa Ana, California)	66	Century 21, real-estate franchisor, represents Bartlett's initial entrepreneurial venture. He sold it to Trans World Corporation. His second is Mr. Build International, a franchised chain of contractors who specialize in residential and commercial remodeling.
Ralph M. Baruch (New York City, New York)	77	A naturalized American citizen born in Germany, Baruch is a force in the cable industry through Viacom International, a systems operator as well as programmer.

Charlie C. Bass, Ph.D. (Santa Clara, California)	58	Bass, a software genius, was tapped by seasoned entrepreneur Ralph Ungermann to cofound Ungermann-Bass, Inc., a maker of general-purpose baseband and broadband local area networks. It's considered one of the finest young firms in the embryonic data communications industry.
C. Andrew L. Bassett, M.D. (Fairfield, New Jersey)	75	His company, Electro-Biology, manufactures sophisticated medical equipment that electromagnetically treats bone fractures.
Jack M. Berdy (Fort Lee, New Jersey)	54	He grew up on the streets of the South Bronx and dreamed about getting rich someday. He may—if he can curb his urge to gamble away the millions he's already made as the founder of On-Line Software International.
N. Edward Berg (Bedford, New Hampshire)	66	While heading up a consulting firm specializing in the graphics industry, he spawned the idea for Bedford Computer Corporation, which manufactures advanced typographical composition systems.
William R. Berkley (Cos Cob, Connecticut)	54	He is the force behind W. R. Berkley Corp., an insurance holding company, and DFI, Inc., a private holding company that owns Dellwood Dairy.
Joel A. Billings (Mountain View, California)	43	Billings has turned his passion for war and history into a thriving business. Strategic Simulations makes video war games for microcomputers and, more recently, sci-fi fantasy games.
Robert Blauth (North Billerica, Massachusetts)	57	Via Systems, the construct of Blauth and three others, makes CAD/CAM equipment for VLSI circuit design.

TABLE 9 *(Continued)*

Candidate (Principal Residence)	Age in Year 2000	Source or Sources of Wealth to Date
David Blech Isaac Blech (New York City, New York)	43 49	These two brothers are promoters involved in the biotechnology industry. Ex-adman and stockbroker, respectively, the brothers raised the money and found the scientist to start Genetic Systems. They repeated their success with such follow-up ventures as DNA Plant Technology Corporation.
Brian Blume Kevin Blume (Lake Geneva, Wisconsin)	50 49	This brother duo and Gary Gygax founded TSR Hobbies Inc., a fast-growing producer of conventional games, toys, and children's books. They publish the immensely popular "Dungeons & Dragons" fantasy role-playing game.
Ivan F. Boesky (New York City, New York)	63	Boesky assembled his fortune as a Wall Street arbitrager.
C. Michael Bowen (Allen, Texas)	61	(See Thomas H. Aschenbrenner)
William Bowman (Cambridge, Massachusetts)	51	With venture capital from TA Associates, Bowman and a fellow Harvard Business School graduate, David Seuss, formed Spinnaker Software to market educational micro packages.
Herbert W. Boyer, Ph.D. (South San Francisco, California)	64	In 1973, Boyer and Stanley Cohen gave birth to the biotechnology industry when they described their new method for recombining fragments of DNA. Three years later, Boyer teamed up with

		businessman Robert Swanson to found Genentech, the leading firm in the industry.
Terry Bradley (Sacramento, California)	not available	He heads Sirius Software, publisher of video games software for microcomputers.
Walter Braun (Englewood, Colorado)	71	Braun, a former science writer, helped design a device that monitors surgery patients' blood and calculates how much coagulant or anticoagulant they need. To market the machine, he formed HemoTec, a fast-growing company.
Donald L. Bren (Orange County, California)	68	He's developing the Irvine Ranch property, which encompasses roughly one fifth of Orange County in sunny Southern California. His mother is actress Claire Trevor.
Robert Brennan (New York City, New York)	56	If you don't recognize the name, you'd recognize the face. He exhorts people to "come grow with us" in the TV ads of his company, First Jersey Securities. It's a brokerage and underwriting firm specializing in high-risk start-up ventures. Racehorses and racetracks are his hobbies.
Daniel Bricklin (Wellesley, Massachusetts)	49	With his best friend, Robert Frankston, he formed Software Arts to create VisiCalc. It's one of the best-selling microsoftware programs of all time. Now Software Arts and VisiCorp, the firm that licensed VisiCalc, are embroiled in a lawsuit over the rights to the famous spreadsheet program.
Dolph Briscoe, Jr. (Uvalde, Texas)	77	His wealth emanates from southwestern Texas real estate, ranching, and local banking. He's a former Texas governor.

TABLE 9 *(Continued)*

Candidate (Principal Residence)	Age in Year 2000	Source or Sources of Wealth to Date
Eli Broad (Los Angeles, California)	67	He's a cofounder of the homebuilding firm of Kaufman & Broad.
Robert Brock (Irving, Texas)	75	Brock made his first millions as the largest domestic franchisee of Holiday Inns. His latest venture is a franchised chain of family restaurants, ShowBiz Pizza Place, featuring robots and video games.
Peter A. Brooke (Boston, Massachusetts)	not available	Brooke is the managing partner of TA Associates, one of the most active and successful venture-capital firms.
Jack Brown (Midland, Texas)	74	Brown teamed up with Cyril Wagner in 1962 to drill for oil. Their luck and profits were so astounding that they set up Canyon, Inc., to invest in other businesses.
Bill Budge (Piedmont, California)	45	Under the aegis of his BudgeCo., his video games are among the most popular, especially with Apple freaks. He's aligned himself with other gifted programmers in a new co-op venture called Electronic Arts.
Warren E. Buffett (Omaha, Nebraska)	70	This premier stock-market investor maintains his distance from the pollution of Wall Street gossip. He says it's a major reason why he makes wise investment decisions.

Donald Burr (Newark, New Jersey)	59	With People Express Airlines he's done the impossible—make money in the airline business at a time when the established carriers are going bust.
Nolan K. Bushnell (Sunnyvale, California)	57	His brain teems with innovative electronic entertainment ideas. The ones he's already implemented—Atari, Pizza Time Theatre—have made him rich. He promises to commercialize more fun and games in the future through his Catalyst Technologies, a company that serves as an incubator for his many start-up ventures.
Jerry Buss (Los Angeles, California)	67	The flamboyant Buss, who holds a Ph.D. in chemistry, is copartner in the lucrative Mariani-Buss Associates (real estate). He owns two Los Angeles teams—the Lakers (basketball) and the Kings (hockey); and he's a 15 percent owner in the Forum arena where both teams play.
Brook H. Byers (San Francisco, California)	55	As a general partner of Kleiner, Perkins, Caufield & Byers (venture capital), he organized, arranged for the financing, and managed during its start-up phase Hybritech, Inc. This successful San Diego firm specializes in medical research and diagnostic products incorporating monoclonal antibodies created by means of "cell fusions."
Ronald E. Cape, Ph.D. (Berkeley, California)	68	He's one of Cetus Corporation's founding trio. Cetus is the oldest firm devoted solely to biotechnology R&D.
Douglas Carlston Gary Carlston (San Rafael, California)	53 49	The brothers head up Broderbund Software, one of the three largest microsoftware game publishers. The privately held company is 70 percent family-owned.

TABLE 9 *(Continued)*

Candidate (Principal Residence)	Age in Year 2000	Source or Sources of Wealth to Date
Johnny Carson (Los Angeles, California)	75	This entertainer is the longtime host of NBC's *Tonight Show.* But his Carson Productions, Inc., a diversified private investment company, is the main source of his wealth.
Richard A. Cerny (Worcester, Massachusetts)	53	He's cofounder of Artel Communications Corporation, a manufacturer of fiber optics transmission systems for analog applications such as broadcasting and teleconferencing.
David Chase (Hartford, Connecticut)	not available	Chase Enterprises develops real estate, particularly shopping centers. Roger Freedman, Chase's son-in-law, is the No. 2 man in the company and shares in Chase's good fortune.
Wendell Cherry (Louisville, Kentucky)	65	With a partner, he started in the nursing home business and branched out into hospitals. Humana, Inc., is now one of the largest private hospital chains in the country.
Otto A. Clark (Melrose Park, Illinois)	74	This Czech immigrant, now a U.S. citizen, is building his fortune in the copier business. Clark Copier International is his third copier company. He holds 55 percent equity in it.
Marshall S. Cogan (New York City, New York)	62	In 1973, this Wall Street operator teamed up with a peer, Stephen C. Swid, to acquire poorly managed companies and turn them around (e.g., General Felt, Kroll International).

Fred Cohen (Moorestown, New Jersey)	73	TeleSciences, his firm, makes computerized telecommunications equipment sold to telephone companies for internal use.
John B. Coleman (Chicago, Illinois)	65	He's off to a good start as a premier hotelier.
James A. Collins (Los Angeles, California)	73	He used his first success, Collins Foods International, to acquire and ultimately spin off his second, Sizzler Restaurants International, a franchised fast-food chain concentrated in the West and Southwest.
Finis F. Conner (Scotts Valley, California)	57	(See Alan F. Shugart.)
Wilfred J. Corrigan (Milpitas, California)	62	LSI Logic Corporation, which he cofounded, sells its services as well as its custom semiconductor VSLI logic circuits, which are designed—often by the customer—using a proprietary computer-based system.
James S. Cownie (Des Moines, Iowa)	56	With James Hoak, he formed Heritage Communications, an operator of cable TV systems. Heritage's secondary business is display advertising.
Edwin L. Cox (Dallas, Texas)	78	Oil made him rich.
John L. Cox (Midland, Texas)	75	This petroleum engineer struck out on his own in 1952. He's now among the largest private oil producers in Texas.

TABLE 9 *(Continued)*

Candidate (Principal Residence)	Age in Year 2000	Source or Sources of Wealth to Date
Seymour R. Cray (Mendota Heights, Minnesota)	74	Cray left Control Data to found a competing firm, Cray Research, which makes high-speed mainframe scientific computers.
John J. Cullinane (Westwood, Massachusetts)	65	His Cullinet Software, Inc., made its reputation by developing integrated data-base management programs for IBM mainframes. It's since diversified, selling a broad range of applications packages.
Frank W. Dabby, Ph.D. (Van Nuys, California)	57	Dabby and his mother, Norma Dabby, own a major interest in Lightwave Technologies, a producer of optical fibers.
Ashraf M. Dahod (Wakefield, Massachusetts)	49	This electrical engineer/M.B.A. started Applitek Corporation to make the hardware and software for local area networks.
Raymond Damadian, M.D. (Melville, New York)	not available	Fonar Corporation markets his invention, a nuclear magnetic resonance scanner for diagnostic imaging.
Barrie M. Damson (New York City, New York)	64	He started in the oil and gas game as a young man, leaving him plenty of time to build a prosperous fiefdom. Lately, commercial real estate has been turning his head.
William M. Davidson (Novi, Michigan)	77	His company, Guardian Industries, makes auto and architectural glass.

Edward M. Davis (Houston, Texas)	67	He's an executive with Tiger Oil International (exploration and production), now a division of Cleveland-Cliffs Iron.
Kenneth DeBower (Austin, Texas)	54	This former computer programmer for Texas Instruments is a major stockholder in BPI Systems, Inc., a producer of business applications software sold or licensed directly to microcomputer manufacturers, especially Apple.
Edson D. deCastro (Westboro, Massachusettes)	62	DeCastro and two partners founded Data General Corporation, one of the early minicomputer manufacturers.
Gary Demos (Los Angeles, California)	50	With another computer graphics savant, (John Whitney, Jr.), he formed Digital Productions. The firm specializes in computer-simulated special effects and digital scene production for the film industry.
Richard M. DeVos (Ada, Michigan)	74	With Jay Van Andel, he formed the Amway Corp. to market household products and cosmetics the Avon way, door-to-door.
Royce Diener (Beverly Hills, California)	not available	Diener's American Medical International is one of the largest private hospital chains in the United States.
C. Norman Dion (Santa Clara, California)	69	Norm Dion uses the profits he realized as a cofounder of Dysan (magnetic memory disks) to back other successful Silicon Valley high-tech start-ups.
Frank Dodge (Needham Heights, Massachusetts)	63	He's the Dodge of McCormack & Dodge, the respected producer of accounting and financial applications software systems, mostly for use on mainframes and minicomputers.

TABLE 9 *(Continued)*

Candidate (Principal Residence)	Age in Year 2000	Source or Sources of Wealth to Date
Charles F. Dolan (Woodbury, New York)	73	His Cablevision Systems Development Corporation is one of the country's most aggressive and innovative multisystem cable TV operators.
Edward R. Downe, Jr. (New York City, New York)	67	With the mail-order business as his base, Downe amassed a fortune through judicious acquisitions in magazine publishing.
Holly Thomis Doyle, Ph.D. Robert O. Doyle, Ph.D. (Cambridge, Massachusetts)	64 64	Despite Ph.D.s in the rarefied field of astrophysics, this husband-and-wife team turned to invention. Their first big hit was Merlin, a toy electronic game. Now, through their own company, IXO, Inc., they're selling a pocket-size computer terminal for adults.
Jerome Drexler (Mountain View, California)	72	His Drexler Technology Corporation holds a patent on an optical disc ideal for data storage. Drexler incorporated this technology into its wallet-size Drexon Laser Card, which stores an unbelievable seven million bits of information, text about as long as this book. It's encoded and read by means of a laser beam.
Richard K. Eamer (Los Angeles, California)	72	National Medical Enterprises, Eamer's company, encompasses nursing homes and hospitals.
Clint Eastwood (Los Angeles, California)	69	Eastwood, the tough loner, pulls them in at the box office with movies that he owns, directs, and stars in. *Dirty Harry, Magnum Force,* and *Escape from Alcatraz* are a few of his better-known flicks.

James J. Edgette (McLean, Virginia)	59	With Steven Heller, he founded Entre Computer Centers, a franchiser of computer stores serving the business market.
Thomas H. Edmondson (North Billerica, Massachusetts)	56	(See Robert Blauth.)
Karl Eller (Phoenix, Arizona)	71	Early in his career, all forms of media were his bailiwick. Eller built Combined Communications and sold it to the Gannett chain for $372 million. Lately he's been disporting himself as a quick-fix, turnaround manager and bouncing around in the broadcasting, film, and convenience store industries.
Carole B. Ely (Thousand Oaks, California)	61	Carole Ely was a housewife when she collaborated with Lore Harp to found Vector Graphic, a maker of business microcomputer systems. Today she's no longer involved with Vector but vows to stage a comeback with a new entrepreneurial venture someday soon.
Heinz Eppler (Secaucus, New Jersey)	71	He heads the retail apparel chain of Miller-Wohl Co.
David C. Evans, Ph.D. (Salt Lake City, Utah)	77	Evans & Sutherland Computer Corporation, which he still heads, is a pioneering manufacturer of three-dimensional interactive graphics systems used for pilot simulation training and CAD/CAM applications.
Anthony J. Faras, M.D. (Minnetonka, Minnesota)	57	He and another physician, Franklin Pass, started Molecular Genetics in 1979. This biotechnology firm develops agriculture and veterinary medicine products.

TABLE 9 *(Continued)*

Candidate (Principal Residence)	Age in Year 2000	Source or Sources of Wealth to Date
Harold Farb (Houston, Texas)	77	From a few buildings constructed with borrowed money in the 1950s, Farb's real-estate-development business has mushroomed into one of leviathan magnitude. His backing of *Ultra,* the Texas version of *Town & Country,* is just for fun.
Peter J. Farley, M.D. (Berkeley, California)	60	A founder of the gene-splicing company Cetus, Dr. Farley resigned in 1983 to form a new company. This company's specialized software helps insurance companies measure the quality and cost efficiency of medical care.
Aaron Fechter (Orlando, Florida)	46	Creative Engineering, this young inventor's company, makes animated robots and attractions for amusement parks. Fechter provides the robots for and is part owner of ShowBiz Pizza Place (see Robert Brock).
Alan Ferry (North Billerica, Massachusetts)	60	(See Robert Blauth.)
Debra Fields Randall Fields (Palo Alto, California)	43 52	Several superb cookie recipes exploded into Mrs. Fields Cookies, Inc. Now Debra Fields's baby is a subsidiary of the private financial services firm of her husband, Randall, Fields Financial Corp., headquartered in Park City, Utah.

Aryeh Finegold (Sunnyvale, California)	53	This Israeli immigrant joined forces with several other engineers to start Daisy Systems Corporation, a manufacturer of computer-aided engineering work stations used in the design of VLSI circuits.
Thomas J. Flatley (Boston, Massachusetts)	68	He's an enormously successful real-estate developer.
Jane Fonda (Los Angeles, California)	62	Everything this crusading actress touches seems to turn to gold. First it was a movie company, IPC (or "*I*ndochina *P*eace *C*ampaign"), which produced four smash hits—all message films, natch—and only one flop. Next came the exercise studios, books, and videocassettes. Now there's a line of versatile workout clothes.
William E. Foster (Natick, Massachusetts)	56	With two partners, he founded Stratus Computer to manufacture a fail-safe minicomputer system.
Arthur L. Fox (Burlington, Massachusetts)	55	He's the cofounder of Octek, Inc., producing artificial-vision systems for robotic applications. Earlier he was a founder of Lexidata, a raster-scan computer graphics company.
Robert M. Frankston (Wellesley, Massachusetts)	51	(See Daniel Bricklin.)
Robert A. Freiburghouse (Natick, Massachusetts)	58	(See William E. Foster.)
Thomas F. Frist, Jr., M.D. (Nashville, Tennessee)	62	He's the youngest member of the trio who founded Hospital Corporation of America, the largest owner and operator of for-profit hospitals.

TABLE 9 *(Continued)*

Candidate (Principal Residence)	Age in Year 2000	Source or Sources of Wealth to Date
Phillip Frost, M.D. (Miami, Florida)	63	Together with Michael Jaharis, Frost owns a controlling stake in the fast-growing Key Pharmaceuticals. In addition to his corporate duties, he still devotes half his time to practicing medicine.
Daniel H. Fylstra (Sunnyvale, California)	49	As a newly minted Harvard M.B.A., he formed VisiCorp with a Canadian, Peter Jennings. One of the first microsoftware packages they licensed to sell, VisiCalc, put their company and Apple Computer on the map. But a latter-day legal dispute over the rights to that same program put the company's future in jeopardy.
Francesco Galesi (New York City, New York)	70	This industrial/commercial real-estate entrepreneur is the catalyst who brought together the disparate forces comprising United Satellite Communications, a direct-broadcast satellite venture. He used the satellite transponder leases held by two young cable programmers as the basis for the joint venture that now includes General Instrument, a TV equipment maker, and insurance giant Prudential. Argo Communications, a satellite-based long-distance telephone company, is another venture.
Janet Gallaher John K. Gallaher (Winston-Salem, North Carolina)	48 48	This married couple is now involved in their second robotics venture. They sold their first company, American Robot Corporation, in 1981. Their current firm, Gallaher Enterprises, makes artificial-vision systems for the small-robot market.

Michael N. Garin (New York City, New York)	54	With Michael J. Solomon, he founded Telepictures Corporation, a fast-growing syndicator and distributor of films and TV programs.
Harry T. Garland, Ph.D. (Mountain View, California)	53	His company, Cromemco, manufactures microcomputers for the high end of the market.
Robert A. Garrow (Santa Clara, California)	57	He's one of the group who founded the enormously successful Convergent Technologies. The company makes a microcomputer-based, multifunctional work station that can be used as a stand-alone system or within a local area network. Convergent's products are sold mostly to original-equipment manufacturers.
Samuel Gary (Denver, Colorado)	73	After thirty-five dry holes, his forty thousand acres of leased land finally gushed "black gold."
William H. Gates (Bellevue, Washington)	44	(See Paul G. Allen.)
David Geffen (New York City, New York)	57	This entertainment tycoon appears incapable of making a bad judgment call when it comes to records, movies, and Broadway shows. Real estate poses his latest challenge.
H. Joseph Gerber (South Windsor, Connecticut)	76	Gerber, an Austrian immigrant, founded Gerber Scientific Instrument Company, which makes CAD/CAM equipment for the apparel, metalworking/plastics, electronics, and graphic arts industries.

TABLE 9 *(Continued)*

Candidate (Principal Residence)	Age in Year 2000	Source or Sources of Wealth to Date
Walter Gilbert, Ph.D. (Cambridge, Massachusetts)	67	This ex-Harvard molecular biologist and Nobel laureate now heads Biogen N.V., one of the four top bioengineering firms in the world. The company is headquartered in Geneva, Switzerland, despite its American financing.
Allen Gilliland, Jr. (San Jose, California)	not available	His father built a tiny empire around Sunlite Bakery, which Gilliland, Jr., sold in the early 1960s to acquire cable systems. The privately held Gill Industries now specializes in cable operation and programming as well as offering on-line data-processing services to the industry.
Donald A. Glaser, Ph.D. (Berkeley, California)	74	This Nobel Prize winner is one of Cetus's founding trio. (See Ronald E. Cape and Peter J. Farley.)
J. Leslie Glick, Ph.D. (Potomac, Maryland)	60	Glick and businessman Robert F. Johnston founded Genex Corporation, a biotechnology firm.
John D. Goeken (Washington, D.C./Joliet, Illinois)	69	This founder of MCI Communications Corporation has long since moved on to other telecommunications ventures. Goeken's latest is AirFone, which offers radiotelephone service from airplanes in flight.
Stanley C. Golder (Chicago, Illinois)	70	Golder heads Golder Thoma & Co., one of the country's premier venture-capital firms. Among the firm's winners are VLI Corporation, maker of a contraceptive sponge.

Lawrence A. Goshorn (Carlsbad, California)	not available	His International Robomation/Intelligence IRI makes robots. Earlier he founded the minicomputer maker General Automation.
Jay Gottlieb (New York City, New York)	55	Jay Gottlieb holds majority equity in The Computer Factory, a fast-growing chain of company-owned retail stores.
William M. Graves (Atlanta, Georgia)	63	He's the only founder who is still a manager of Management Science America. (See John P. Imlay, Jr.)
Michael Green (Cupertino, California)	56	Green is one of Tandem Computers' founding quartet. (See James G. Treybig.)
Maurice R. Greenberg (New York City, New York)	75	Greenberg built the American International Group into the leading insurance holding company it is today.
Richard E. Greene (Norwalk, Connecticut)	62	His Data Switch Corporation makes a computer/peripheral switch for IBM multiple mainframe installations.
Frederick A. Gross (Malvern, Pennsylvania)	61	Gross is the largest individual stockholder of Systems & Computer Technology Corporation, which markets applications software and data-processing services to institutions of higher education and government agencies.
Bud Grossman (Eden Prairie, Minnesota)	78	His Gelco Corp. is a privately held vehicle for soft-drink bottling, transportation leasing, and petroleum marketing.

TABLE 9 *(Continued)*

Candidate (Principal Residence)	Age in Year 2000	Source or Sources of Wealth to Date
Robert C. J. E. S. (Bob) Guccione (New York City, New York)	68	*Penthouse,* the racy girlie magazine, is Guccione's answer to *Playboy.* Try as he might, Guccione still can't match Hugh Hefner when it comes to eccentricity and high living, even though his balance sheet long since put Hef to shame.
E. Gary Gygax (Beverly Hills, California)	61	A former insurance underwriter, he's a founder of TSR Hobbies, Inc. "TSR" stands for *T*actical *S*tudies *R*ules. (See Brian Blume and Kevin Blume.)
William Hambrecht (San Francisco, California)	64	In 1968 he and a partner, who died in 1982, founded Hambrecht & Quist, Inc., an investment banking firm.
Anthony Hamilton (New York City, New York)	76	Hamilton founded a firm purchased by Avnet, Inc., the world's largest distributor of electronic components to industry and the military. Now he is Avnet's chairman.
Floyd R. Hardesty (Tulsa, Oklahoma)	62	Hardesty's formula for success in residential and commercial real estate is: Build it simple, cheap, and fast.
Lore Harp (Thousand Oaks, California)	66	(See Carole B. Ely.) Divorced from one man on this list—Bob Harp—she's remarried to another—Pat McGovern. Today she heads up a San Francisco-based venture-capital fund that invests half its capital in young Japanese companies.

Robert S. Harp, Ph.D. (Westlake Village, California)	63	This engineer/physicist designed the circuit board that gave his ex-wife Lore Harp and Carole Ely a foothold in the microcomputer business back in 1976. On his own, Dr. Harp founded Corona Data Systems in 1981. The company makes IBM PC-compatible microcomputers.
Dennis C. Hayes (Norcross, Georgia)	50	Hayes, an engineer, jumped into the market for microcomputer modems so early that his Hayes Microcomputer Products now holds the predominant market share.
Steven B. Heller (McLean, Virginia)	60	(See James J. Edgette.)
Gardner C. Hendrie (Natick, Massachusetts)	68	(See William E. Foster.)
John Hendrix (Midland, Texas)	64	This petroleum engineer struck out on his own at age thirty and struck oil soon thereafter. He also has banking and timber interests in Arkansas.
Thomas A. Hiatt (Palo Alto, California)	52	He worked for International Plant Research Institute before cofounding a competing company in 1981. That company, Sungene Technologies, specializes in the genetic modification and improvement of major field crops.
J. Donald Hill, M.D. (Berkeley, California)	68	This cardiovascular surgeon teamed up with Robert J. Harvey, a biomedical engineer, to found Thoratec Laboratories. It sells artificial hearts as well as other medical equipment.

TABLE 9 *(Continued)*

Candidate (Principal Residence)	Age in Year 2000	Source or Sources of Wealth to Date
Gary Hillman (Whippany, New Jersey)	61	His Machine Technology, Inc., serves the semiconductor industry as a maker of silicon wafer-processing equipment.
Gerald D. Hines (Houston, Texas)	75	Hines started small in the real-estate business, undertaking increasingly large-scale projects. He built Shell Oil's fifty-story headquarters and went nationwide in 1973.
Neil S. Hirsch (New York City, New York)	52	His Telerate terminals are found on desktops throughout the financial community because his system reports the prices of almost every instrument traded on Wall Street. Hirsch formed his electronic data service in 1969.
James M. Hoak, Jr. (Des Moines, Iowa)	56	(See James S. Cownie.)
Bernard C. Hogan (Dallas, Texas)	73	A cofounder of Hogan Systems (mainframe financial software) with Gregor Peterson, he left just before the company went public. He's now involved in high-technology telecommunications projects.
Albert L. Horley (Mountain View, California)	64	This physicist founded Vitalink Communications Corporation to sell satellite and earth data-transmission equipment and service to American companies.
Robert S. Howard (Rancho Santa Fe, California)	75	This ex-journalist bought his first newspaper in 1961. Today Howard Publications owns nearly twenty dailies and is diversifying into radio-TV station ownership.

Don Hudgins (Norcross, Georgia)	55	With three partners—all with equal equity—he owns Cinetron Computer Systems, which makes computer control systems used in animation and special-effects photography. Cinetron also offers computer-graphics services.
K. P. (Phil) Hwang (Sunnyvale, California)	63	This Korean came to the United States as an engineering student and never went home again—except to arrange for Korean laborers to assemble his company's products. TeleVideo Systems sells high-resolution, microprocessor-based video display terminals. Its latest offering is a multi-user microcomputer system.
John P. Imlay, Jr. (Atlanta, Georgia)	64	Imlay used his adroit management skills to salvage Management Science America, which is now the country's largest independent supplier of off-the-shelf software packages for all sizes of computers. His reward is stock representing almost 25 percent equity.
David Jackson (San Jose, California)	63	This Englishman came to the United States in the early 1960s. By 1971 he'd already founded and sold off his first company, Peripheral Technology. His next, Altos Computer Systems, makes both eight- and sixteen-bit microcomputers sold to original-equipment manufacturers.
Irwin Jacobs (Minneapolis, Minnesota)	58	Jacobs is an aggressive investor who takes controlling positions in well-known American companies. Jacobs Industries, Inc., is his private holding company.
Raymond E. Jaeger, Ph.D. (Sturbridge, Massachusetts)	62	(See Mohd A. Aslami.)

TABLE 9 *(Continued)*

Candidate (Principal Residence)	Age in Year 2000	Source or Sources of Wealth to Date
Michael Jaharis, Jr. (Miami, Florida)	72	(See Phillip Frost.)
Robert K. Jarvik, M.D. (Salt Lake City, Utah)	54	Kolff Medical, Inc., which Jarvik cofounded with several other doctors, markets artificial organs. Among them is the JARVIK-7, the first permanent and totally synthetic heart ever implanted in a human patient (1982).
Boyd Jefferies (Los Angeles, California)	69	He started his brokerage firm in 1962, sold it to Investors Diversified Services, then bought it back four years later. Jefferies & Co. specializes in block trading for institutions but plans to move into retail trading.
Peter Jennings (Sunnyvale, California)	not available	(See Daniel H. Fylstra.)
Richard Jennings (North Billerica, Massachusetts)	62	(See Robert Blauth.)
Steven P. Jobs (Cupertino, California)	45	With Steve Wozniak, a high-school friend, and businessman Mike Markkula, he started Apple Computer. It was one of the first companies to market a fully assembled microcomputer.
Robert F. Johnston (Princeton, New Jersey)	63	(See J. Leslie Glick.)

David A. Jones (Louisville, Kentucky)	69	(See Wendell Cherry.)
Glenn R. Jones (Denver, Colorado)	not available	His Jones Intercable is the holding company for several cable TV systems.
Thomas Jordan (Healdsburg, California)	65	This geologist/attorney sold his Trend Exploration Ltd. (oil) in 1974 and used some of the proceeds to set himself up as a gentleman vintner in Sonoma County, California.
Mitchell D. Kapor (Cambridge, Massachusetts)	50	A superb programmer in his own right, Kapor founded Lotus Development Corporation to design and sell microsoftware. Its first product, 1-2-3, pushed the company to a leading position in the industry.
Charles A. Kappenman (Tinton Falls, New Jersey)	58	He's the founder and was once the largest shareholder in Eagle Computer. Eagle, the maker of a full line of small business and personal desktop microcomputer systems compatible with IBM's PC, ran into financial trouble in 1984 and Kappenman resigned. He went back to his first love—another company he founded, Audio Visual Laboratories.
Harold Katz (Philadelphia, Pennsylvania)	63	Under the respective banners of Nutri/Systems and Cosme-Dontics, Katz operates a chain of weight-loss clinics and a franchised chain of dental offices. Just for fun, he owns the Philadelphia 76ers basketball team.
James Katzman (Cupertino, California)	54	(See James G. Treybig.)

TABLE 9 *(Continued)*

Candidate (Principal Residence)	Age in Year 2000	Source or Sources of Wealth to Date
Thomas S. Kavanagh (Boulder, Colorado)	65	He was a cofounder/manager of Storage Technology when he was recruited in 1975 to head the two-year-old NBI, Inc. Today NBI is one of the leading manufacturers of word-processing equipment.
Joseph F. Keenan (Sunnyvale, California)	59	Keenan is the longtime business partner of Nolan K. Bushnell. Their most famous co-ventures are Atari and Pizza Time Theatre.
Richard Kessler (Atlanta, Georgia)	53	Cecil B. Day, the late founder of Days Inns (motels), was his mentor. After Day's death, Kessler stayed on as the company chairman, garnering a growing percentage of equity in the privately held lodging chain. He cashed in and left when the company was purchased by Reliance Group Holdings in 1984.
Gary Kildall, Ph.D. Dorothy McEwen Kildall (Pacific Grove, California)	58 57	This married couple built Digital Research around the microcomputer operating system (CP/M) that Kildall designed. He's a former computer science professor.
Don King (Las Vegas, Nevada)	68	King runs a passel of athletic and entertainment ventures. They range from fight promotion—he maintains promotional control over almost half of the current world champion boxers—to lucrative pay-per-view cable TV deals. King's distinctive upsweep hairdo makes him look like he's just seen a ghost.

Calvin Klein (New York City, New York)	58	His name is splashed across a high-priced line of men's and women's ready-to-wear clothes, not to mention the millions of pairs of blue jeans sporting his label. In 1983 he and partner Barry Schwartz bought Puritan Fashions, the licensee that markets Klein's clothes.
Eugene Kleiner (San Francisco, California)	76	The Vienna-born Kleiner was one of the founders of Fairchild Semiconductor. Today his name is included on the door of an extremely profitable venture-capital operation called Kleiner, Perkins, Caufield & Byers.
Robert A. Kleist (Irvine, California)	71	He's the founder retaining the greatest equity in Printronix, a producer of dot-matrix low- and medium-speed printers with graphics capability.
Cleon T. Knapp (Los Angeles, California)	62	The privately held Knapp Communications encompasses magazine and book publishing as well as distribution. *Architectural Digest* is the flagship publication.
Philip H. Knight (Beaverton, Oregon)	62	Knight teamed up with a college track coach in 1965 to sell Nike running shoes. Their timing was perfect, since running soon exploded into a national craze.
Sandra Kurtzig (Los Altos, California)	53	Impending motherhood made this aeronautical engineer start a software business in her home in 1972. Ten years later, her ASK Computer Systems was a publicly held company and her 60 percent equity was worth $40 million.

TABLE 9 *(Continued)*

Candidate (Principal Residence)	Age in Year 2000	Source or Sources of Wealth to Date
Henryk de Kwiatkowski (New York City, New York)	75	This European immigrant is a wildly successful used-airplane broker and the lucky owner of Conquistador Cielo, the 1982 Belmont Stakes winner.
C. Kevin Landry (Boston, Massachusetts)	56	Landry is a managing partner of TA Associates, a venture-capital firm whose winners include Tandon and Biogen.
Hal Lashlee (Culver City, California)	not available	His software empire spans publishing (Ashton-Tate Co.), wholesale distribution (Software Distributors), and retailing (Softwaire Centres International). His cofounder, George Tate, died suddenly of a heart attack in 1984. Tate was 40.
Ralph Lauren (New York City, New York)	61	Ralph Lauren's Polo Fashions line of sportswear is the ultimate in casual chic—and very expensive.
M. Larry Lawrence (San Diego, California)	74	Lawrence did his first small building contractor jobs as a teenager during summer vacations from school in Chicago. When his hometown no longer proved a challenge, he moved his phenomenally successful real-estate operation west to California.
Norman Lear (Los Angeles, California)	77	Embassy Communications, which Lear owns with several partners, is a holding company for film, TV, and video enterprises. Many hail him as the country's most innovative TV producer (e.g., *All in the Family, Mary Hartman, Mary Hartman*).

David Lee (San Jose, California)	62	Born in Peking and raised in South America, Lee came to the United States to make his fortune. He founded Qume Corporation, a manufacturer of high-speed, letter-quality, daisy-wheel printers. Qume was purchased by ITT in 1978 for $127 million in stock.
Robert Leff (Los Angeles, California)	53	With David Wagman, he built Softsel Computer Products into the prime microsoftware distributor it is today.
Gerald A. Lembas (Chatsworth, California)	61	He had the good fortune to join Tandon Corporation (disk drives) during its start-up phase. The stock he received as compensation in those early days eventually made him a millionaire many times over.
Mark Leslie (Milpitas, California)	54	Leslie and Elliot Nestle cofounded Synapse Computer Corporation, which makes a fault-tolerant mainframe computer for transaction processing.
James H. Levy (Milpitas, California)	55	This former record-industry executive spearheaded the founding of Activision, the video games software publisher.
Lawrence F. Levy Mark Levy (Chicago, Illinois)	56 53	These brothers founded the Levy Organization, a family-owned holding company for delis, restaurants, a supermarket and catering chain, two clothing chains, and a real-estate development division.
Walter Loewenstern, Jr., Ph.D. (Santa Clara, California)	64	He's a member of the quartet who founded ROLM Corporation, a high-flying telecommunications equipment maker.

TABLE 9 *(Continued)*

Candidate (Principal Residence)	Age in Year 2000	Source or Sources of Wealth to Date
Joseph Looney, Jr. (Tampa, Florida)	not available	He founded Paradyne in 1969, then lost control to its investors. Now he's trying again with Vistar Corporation, a producer of "wave analyzers"—i.e., instruments to test and measure complicated electrical signals. This time Looney and his partner, Tom Saliga, control 50 percent of the stock.
George Lucas (San Rafael, California)	56	Through Lucasfilm Ltd., this producer-writer-director owns all rights to *Star Wars* and its sequels. Lucasfilm is also a full-service movie studio outfitted with the most advanced filmmaking equipment.
Betsy Magness Robert Magness (Denver, Colorado)	76 76	They nurtured their Tele-Communications, Inc., from a tiny mom-and-pop cable company into the gigantic multisystem operator/owner it is today. In addition, it serves other telecommunications companies through the microwave relay system it owns.
John N. Maguire (Reston, Virginia)	70	Maguire spun off Software AG International from a West German company and still retains more than 25 percent equity. The company sells packaged software for medium-to-large-scale IBM computers.
E. Blaine Mansfield, Jr. (Signal Hill, California)	58	AMFOX, a producer of optical fibers, is his third entrepreneurial venture. His first two—Microtech Data Systems and Microbyte—developed computer hardware and software products.

Frank Mariani (Los Angeles, California)	64	(See Jerry Buss.)
A. C. (Mike) Markkula, Jr. (Cupertino, California)	58	(See Steven P. Jobs.)
Cloyd Marvin (Santa Clara, California)	64	He's a founder of Four-Phase Systems and Compression Labs. Now he's a venture capitalist associated with Harvest Ventures.
John Masefield (Whippany, New Jersey)	66	A mechanical engineer by training, Masefield founded Isomedix, Inc., the world's largest independent irradiation service company. (Irradiation involves the application of gamma rays to commercial products to stop cell growth, among other beneficial effects.)
Irwin Math (Port Washington, New York)	59	With his wife, he started a mail-order business specializing in fiber optics components. Today, Math Associates makes fiber optics data-transmission systems for industrial applications.
Gordon Matthews (Richardson, Texas)	63	He's a telecommunications inventor of the first order. His business skills are less well developed, though, causing him to lose control of his first two companies. His current company, VMX, Inc., designs and sells a computerized telephone message storage and retrieval system.
Robert R. Maxfield, Ph.D. (Santa Clara, California)	58	(See Walter Loewenstern, Jr.)
Billy J. (Red) McCombs (San Antonio, Texas)	72	He believes, like Dolly Levi, that money and manure are for spreading around. His holdings range from convenience stores, auto dealerships, and radio and TV stations to oil wells, airlines, and ranching.

TABLE 9 *(Continued)*

Candidate (Principal Residence)	Age in Year 2000	Source or Sources of Wealth to Date
James McCormack (Needham Heights, Massachusetts)	63	(See Frank Dodge.)
Patrick J. McGovern (Nashua, New Hampshire)	63	International Data Group, which he started in 1964, has grown into the world's largest provider of information about computers, office systems, and telecommunications. Among its many publications is *Computerworld.* In 1982, McGovern married Lore Harp, cofounder of Vector Graphic.
William G. McGowan (Washington, D.C.)	73	After founding and selling off his own company (Ultrasonic), he became a consultant aiding other people's stalled ventures. One of them was MCI Communications Corporation, which he built into a company that today is second only to AT&T in the long-distance telephone service business.
Regis McKenna (San Francisco, California)	61	The public relations firm bearing his name specializes in the promotion of high-technology start-up firms. PR services alone probably will never make him superrich. But his venture-capital investments in the types of companies he publicizes may. He's also been known to barter services for stock.
John J. McNaughton (Newport Beach, California)	not available	McNaughton is the founder of National Education Corporation, the largest chain of vocational and correspondence schools in the country.

Lorraine Mecca (Fountain Valley, California)	51	It took this restless woman a while to settle on an occupation. She tried teaching school, receptionist, and fashion show producer. She finally found her niche as an entrepreneur heading up her own Micro D. Inc., a wholesaler of software and microcomputer peripherals.
John Meehan Paula Meehan (Los Angeles, California)	71 68	This couple founded Redken Laboratories, makers of hair and skin-care products.
Roger D. Melen, Ph.D. (Mountain View, California)	53	(See Harry T. Garland.)
Robert M. Metcalfe, Ph.D. (Mountain View, California)	54	As an employee of Xerox, he was a designer of Ethernet. He left to cofound 3Com Corporation, a manufacturer of software and hardware interconnect equipment for building local area networks.
Paul J. Meyer (Waco, Texas)	72	His Success Motivation Institute (SMI) peddles self-improvement courses and products, among its other diversified interests. About half of his swelling net worth is due to savvy real-estate investments.
Allen H. Michels (Santa Clara, California)	59	(See Robert A. Garrow.)

TABLE 9 *(Continued)*

Candidate (Principal Residence)	Age in Year 2000	Source or Sources of Wealth to Date
William H. Millard (Hayward, California)	68	In 1975, Millard founded IMS Associates, which remains the personal holding company for his diverse business ventures. His manufacturing company, called IMSAI, sold some of the first computer kits and, later, fully assembled microcomputers. That company failed, but Millard's ComputerLand—he owns 94 percent equity—is worth a mint. It's the largest privately held franchised chain of its kind.
Howard R. Miller (Allen, Texas)	53	(See Thomas H. Aschenbrenner.)
Ken Miller (Waltham, Massachusetts)	60	He left Codex, a market leader in the sale of modems, to found a competing firm, Concord Data Systems. A local area network is on Concord's drafting board.
Rudy Miller (Phoenix, Arizona)	53	In 1973 Miller bought a vocational school that evolved into the successful Miller Technology & Communications, specializing in technical training (electronics, computer science, broadcasting, CAD/CAM); applications software for educational institutions; and video programming.
Phillip S. Mittelman, Ph.D. (Elmsford, New York)	74	Mittelman founded MAGI in 1967, making it one of the earliest providers of products and services in the computer-graphics field. An automated slide production machine is another company money-maker.

Gordon E. Moore, Ph.D. (Santa Clara, California)	71	A cofounder of Fairchild Semiconductor back in 1957, Moore is also one of Intel's founding troika (1968). Intel makes a broad line of innovative semiconductor products and microprocessors.
Mary Tyler Moore (New York City, New York; Los Angeles, California)	63	Born in Brooklyn, this businesswise actress struck a vein of gold in Hollywood. Her MTM Enterprises produces such quality TV fare as *The Bob Newhart Show* and *Hill Street Blues.*
Robert A. Mosbacher (Houston, Texas)	72	This native easterner moved to Texas in 1948 to become an oil wildcatter. The move paid off, netting him, in addition to multitudinous barrels of black goo, hundreds of acres of real estate and a ranching fiefdom.
Gary Mounts (Anaheim, California)	54	With one successful medical equipment company already to his credit, Mounts started Emergent Corporation, the maker of a microprocessor-based diagnostic imaging device.
Charles A. Muench, Jr. (Norcross, Georgia)	63	Intelligent Systems Corporation, his second successful corporate venture, manufactures desktop computer terminals with color graphics capability. PrintaColor, his third company, makes color printers that connect to his terminals and churn out hard copies.
David H. Murdock (Los Angeles, California)	77	This high-school dropout started his entrepreneurial career in Phoenix real estate, moved to California to exploit that boom, then used his profits to invest in undervalued companies. A bulging stock portfolio forms the basis of his fortune today.
W. Edward (Ted) Naugler, Jr. (Signal Hill, California)	61	Naugler sold his first company to M/A-Com in 1980 and immediately cofounded his second, AMFOX. (See E. Blaine Mansfield, Jr.)
Elliot Nestle (Milpitas, California)	65	(See Mark Leslie.)

TABLE 9 *(Continued)*

Candidate (Principal Residence)	Age in Year 2000	Source or Sources of Wealth to Date
Wayne Newton (Las Vegas, Nevada)	57	This popular singer/entertainer has made astute investments in hotels, casinos, and other real estate, as well as in Arabian horses.
David Norman (San Jose, California)	65	Norman founded Dataquest (market research) in 1972 and sold it to the A. C. Nielsen Company six years later for $5 million. With Enzo Torresi, he's now building Businessland, a hybrid chain of computer/office-equipment stores.
William A. Norred (Chatsworth, California)	59	Norred founded the predecessor firm that evolved into Micom Systems, a manufacturer of microcomputer-based data communications gear. Micom's principal product is a data concentrator—sometimes called a "statistical multiplexor"—that enables information from a number of data terminals to be combined for cost-effective transmission over a single communications link.
Gary A. Norton (Spokane, Washington)	not available	ISC System Corporation (microprocessor-based terminals) is one of the fast-growing high-technology companies in the country.
Gene Nottingham (Norcross, Georgia)	57	(See Don Hudgins.)
Robert Nowinski, Ph.D. (Seattle, Washington)	53	Nowinski was the scientist recruited by the Blech brothers (see Isaac Blech and David Blech) to run Genetic Systems, specializing in monoclonal antibody R&D.

Eric Nowlin, Ph.D. (New York City, New York)	54	With a doctorate in communications theory and film, he teamed up with businessman Robert Robbins to found VIDMAX, a pioneering producer of interactive laser videodisks for the educational and entertainment markets.
Robert N. Noyce, Ph.D. (Santa Clara, California)	71	(See Gordon E. Moore.)
Edmund J. O'Connor William F. O'Connor (Chicago, Illinois)	76 70	In 1973, the secretive O'Connor brothers founded First Options of Chicago (clearinghouse for stock options trades), which they sold in 1979. Today they trade for their own account, concentrating on commodities futures, stock options, and risk arbitrage.
Kenneth H. Olsen (Lincoln, Massachusetts)	74	Digital Equipment Corporation (DEC), the company he founded and still heads, is the country's foremost minicomputer manufacturer.
Elisabeth Claiborne Ortenberg (New York City, New York)	71	In 1976 this talented designer lent her name to an elegant yet serviceable line of women's ready-to-wear clothing. Five years later, she and her husband, Arthur, took Liz Claiborne, Inc., public, and their personal net worth has been growing steadily ever since.
M. Kenneth Oshman, Ph.D. (Santa Clara, California)	60	(See Walter Loewenstern, Jr.)
Stanford R. Ovshinsky (Troy, Michigan)	77	This self-educated engineer started Energy Conversion Devices in 1960 to develop special "amorphous materials" to incorporate into energy production equipment such as photovoltaic cells. The company also licenses a laser device used in optical videodisk storage technology.

TABLE 9 *(Continued)*

Candidate (Principal Residence)	Age in Year 2000	Source or Sources of Wealth to Date
David Padwa (Golden, Colorado)	68	He's the classic entrepreneur, willing to go into whatever field offers the most promise of financial reward. His current venture is Agrigenetics, a seed company that plows its profits into food crop R&D utilizing genetic engineering techniques. Padwa sold it to Lubrizol Corp. in 1984.
Max Palevsky (Los Angeles, California)	76	The stock he received in the late 1960s in exchange for his major interest in Scientific Data Systems—the mainframe computer firm he founded—made him for years Xerox's largest shareholder. Now he's an investor.
Victor H. Palmieri (Los Angeles, California)	70	Victor Palmieri & Company offers money management services to institutional investors. It also takes large equity positions in "sick" companies it restores to health. Palmieri is best known for reorganizing Penn Central during its 1970 bankruptcy.
Franklin Pass, M.D. (Minnetonka, Minnesota)	63	(See Anthony J. Faras.)
Allen E. Paulson (Savannah, Georgia)	78	Since 1951 he's been involved in entrepreneurial ventures in the aerospace field. But he didn't amass his Croesus-like fortune until he bought a foundering Grumman division. It evolved into the successful Gulfstream Aerospace Corporation.

Hal Pearson (Norcross, Georgia)	61	(See Don Hudgins.)
A. Jerrold Perenchio (Los Angeles, California)	70	Perenchio made his first multimillion-dollar bundle in 1981, when he sold his 49 percent interest in National Subscription Television. Now he's a principal in Embassy Communications (see Norman Lear).
Thomas J. Perkins (San Francisco, California)	68	He's one of the two original partners in the venture-capital firm that's evolved into Kleiner, Perkins, Caufield & Byers. (See Eugene Kleiner.)
Clifford S. Perlman Stuart Z. Perlman (Los Angeles, California)	74 not available	This brother duo invests in gaming and hotel/resort properties.
H. Ross Perot (Dallas, Texas)	70	Perot is fascinated with all things military and regimented. It shows in the way he runs his Electronic Data Systems, a facilities management data-processing firm he sold to General Motors in 1984. The transaction made him a billionaire in cool, hard cash.
Robert E. Petersen (Beverly Hills, California)	74	With borrowed money he built Petersen Publishing into a thriving, special-interest magazine (e.g., *Hot Rod, Guns & Ammo*) publisher.
Gregor G. Peterson (Petaluma, California)	68	Peterson retains almost a third of the ownership in the Dallas-based company he cofounded—Hogan Systems, a purveyor of financially oriented applications software for IBM mainframes and compatible hardware systems.

TABLE 9 *(Continued)*

Candidate (Principal Residence)	Age in Year 2000	Source or Sources of Wealth to Date
T. Boone Pickens, Jr. (Amarillo, Texas)	72	Pickens is the founder of Mesa Petroleum Company. He's known as a "financial shark" because of his proclivity for organizing unfriendly take-over bids.
Ann Piestrup, Ph.D. (Portola Valley, California)	57	This educational psychologist heads a prosperous microsoftware firm—the Learning Company—specializing in interactive programs for children aged three to thirteen.
Arthur A. Pilla, Ph.D. (Fairfield, New Jersey)	63	(See C. Andrew L. Bassett.)
John William Poduska, Ph.D. (Chelmsford, Massachusetts)	62	Poduska was one of the founders of Prime Computer (minicomputers) in 1971. He liked being an entrepreneur so much that he did it again ten years later. His Apollo Computer makes superminicomputer networking systems used by teams of engineers and scientists.
Kenneth N. Pontikes (Barrington, Illinois)	60	The urge to become his own man was strong in Pontikes when he quit his job as an IBM salesman and launched a used-computer operation. In a little over a decade, he turned his Comdisco, Inc. into the foremost broker of revamped minis and mainframes, mostly IBMs.

Generoso P. Pope, Jr. (Manalapan, Florida)	73	He bought the *National Enquirer* in 1952 for $75,000. Today this weekly tabloid is distributed in supermarkets all over the country and bought by 5.2 million readers eager for celebrity gossip and medical cure-alls.
David A. Post (New York City, New York)	59	Post was a Wall Street research analyst specializing in emerging growth industries when he conceived the notion for his paging and message information network, PageAmerica.
Wilber L. Pritchard (Bethesda, Maryland)	76	He's the largest individual stockholder in Direct Broadcast Satellite Corporation. In the mid-1980s, this common carrier plans to offer direct-broadcast satellite service to homes in remote regions.
William J. Pulte (Detroit, Michigan)	68	His Pulte Home Corporation builds single-family, no-frills houses for the working classes. He's semiretired already and enjoying his wealth.
Leslie Quick, Jr. (New York City, New York)	74	Quick left a management job at *Forbes* magazine to cofound Quick & Reilly, a discount stock-brokerage house.
Elazar Rabbani, Ph.D. Shahram K. Rabbani (New York City, New York)	56 48	With their brother-in-law Barry Weiner, the Rabbani brothers started Enzo Biochem, which purifies, produces, and sells specialized enzymes to research labs engaged in nucleic acid and immunological research. Enzo Biochem also undertakes its own recombinant DNA research.
Eric N. Randall, Ph.D. (Cheshire, Connecticut)	61	His Norrsken Corporation builds fiber optics manufacturing equipment.

TABLE 9 *(Continued)*

Candidate (Principal Residence)	Age in Year 2000	Source or Sources of Wealth to Date
Sheldon Razin (Tustin, California)	62	Razin and his wife own almost 60 percent of his Quality Systems. The firm develops applications software, which it integrates into a minicomputer-based DP system sold to dental group practices.
Sumner Redstone (Newton Center, Massachusetts)	77	This ambitious lawyer made his first killing with a chain of drive-in movies. Now he invests his profits in other segments of the entertainment industry, such as film studios, family restaurants, and video arcades.
Ronald C. Reeve, Jr. (Hebron, Ohio)	57	Reeve negotiated a leveraged buyout of a division of Air Products & Chemicals. The result is Advanced Robotics, specializing in custom arc-welding systems.
Jeffrey C. Reiss (New York City, New York)	58	Born in Brooklyn, Reiss taught in public schools until Norman Lear made him a scriptreader. Then he became a TV titan in his own right as the mile-a-minute developer of the cable network Showtime. He's got more big-money programming ideas, which he plans to turn into cash through his consulting firm.
Herbert J. Richman (Westboro, Massachusetts)	65	(See Edson D. deCastro.)
Monroe M. Rifkin (Denver, Colorado)	69	He left cable broker Daniels & Associates to found American Television & Communications. He sold this cable giant in 1978 to

		Time, Inc., and now is a free-lance wheeler-dealer in the cable industry.
Richard J. Riordan (Santa Clara, California)	69	(See Robert A. Garrow.)
Robert Robbins (Cincinnati, Ohio)	52	(See Eric Nowlin.)
Gordon I. Robertson, Ph.D. (Princeton, New Jersey)	57	He met Anton Rodde at Western Electric. They resigned to found Control Automation, which designs and manufactures integrated robotics systems.
Arthur Rock (San Francisco, California)	74	He's one of the earliest venture capitalists to mine the West Coast high-tech terrain—and today he's probably the richest. His bull's-eye winners include Scientific Data Systems, Teledyne, Apple, and Diasonics.
Anton F. Rodde, Ph.D. (Princeton, New Jersey)	57	(See Gordon I. Robertson.)
Kenny Rogers (Los Angeles, California)	62	This country-style singer/entertainer is a perennial favorite with audiences. His record royalties prove it.

TABLE 9 *(Continued)*

Candidate (Principal Residence)	Age in Year 2000	Source or Sources of Wealth to Date
Richard R. Rogers (Dallas, Texas)	57	Rogers started Mary Kay Cosmetics with his mother, Mary Kay Ash, in 1963. The products caught on due to the conducive sales ambiance at the Tupperware-type parties given by company saleswomen.
Robert M. Rogers (Tyler, Texas)	73	He's the founder/majority stockholder in TCA Cable TV, with twenty-three cable systems located primarily in the smaller markets of Texas, Louisiana, and Arkansas.
Paul F. Rollins (Phoenix, Arizona)	not available	His Phoenix Digital Corporation designs and produces fiber-optics-based local area networks linking all the automated machines, robots, and computers in factories.
William D. Rollnick (Santa Clara, California)	67	(See Robert A. Garrow.)
Judson Rosebush (New York City, New York)	59	Rosebush is the largest founding stockholder in Digital Effects, offering computer animation products/services.
Ben Rosen (New York City, New York)	66	Ben Rosen brings to his newfound role of venture capitalist an impressive lineup of degrees: B.S. in electronics engineering, California Institute of Technology; M.S. in science and electrical engineering, Stanford; M.B.A., Columbia. No wonder he can really pick 'em! (See L. J. Sevin.)

Douglas T. Ross (Waltham, Massachusetts)	70	His SofTech, Inc., provides diversified software consulting services. It is particularly adept at developing state-of-the-art systems software for military weapons systems.
Anthony Cal Rossi, Jr. (San Francisco, California)	64	Rossi is a restaurateur/hotelier/real-estate developer. His showpieces: the Stanford Court Hotel and the Pacific Plaza.
Seymour I. Rubinstein (San Rafael, California)	66	After an unprofitable start-up phase, MicroPro International, a business applications software publisher, is growing fast.
John P. Ryaby (Fairfield, New Jersey)	65	(See C. Andrew L. Bassett.)
Harry J. Saal, Ph.D. (Palo Alto, California)	55	Nestar Systems, his company, makes local area network systems for linking microcomputers, particularly Apples and IBM PCs.
John W. Sammon, Jr., Ph.D. (New Hartford, New York)	61	Sammon controls almost two thirds of the equity in his PAR Technology Corporation. Its product is a microprocessor-based point-of-sale terminal for the fast-food industry. PAR is also a Defense Department contractor.
W. Jerry Sanders III (Sunnyvale, California)	64	Sanders left a marketing post at Fairchild Semiconductor to start a competing firm, Advanced Micro Devices.
Herbert Sandler, Esq. Marion O. Sandler (Oakland, California)	69 70	Marion (B.A. in economics, Wellesley) and Herbert (J.D., Columbia) bought Golden West Financial Corporation in 1963 for $4 million and built it into a large savings-and-loan holding company.
Charles J. Satuloff (White Plains, New York)	71	Satuloff, in conjunction with *The New Yorker* magazine, owns Teleram Communications Corporation, a manufacturer of portable microcomputers, particularly an editing terminal for news reporters filing stories from remote locations.

TABLE 9 *(Continued)*

Candidate (Principal Residence)	Age in Year 2000	Source or Sources of Wealth to Date
Dean F. Scheff (Eden Prairie, Minnesota)	68	His CPT Corporation makes word-processing equipment sold to smaller companies in out-of-the way places.
Charles Schwab (San Francisco, California)	62	He sold Charles Schwab & Co. (discount brokers) in 1981 to BankAmerica for $53 million. His profit: $18.5 million.
Barry K. Schwartz (New York City, New York)	58	This boyhood friend of Calvin Klein, clothing designer from the Bronx, is now Klein's business partner.
Bernard L. Schwartz (New York City, New York)	74	Schwartz is the strategist and troubleshooter who engineered the diversification of Saul Steinberg's empire. Now Schwartz is working similar miracles at Loral Corporation, a maker of state-of-the-art electronic warfare devices for defense and communication.
Michael M. Scott (Cupertino, California)	57	He had the prescience to join Apple Computer as CEO during its early years. Stock options made him rich.
Howard W. Selby III (Boulder, Colorado)	58	Selby founded NBI, the leading word-processor manufacturer. (See Thomas S. Kavanagh.)
David Seuss (Cambridge, Massachusetts)	50	(See William Bowman.)

L. J. Sevin (Dallas, Texas)	69	He founded Mostek (semiconductors), then sold it to United Technologies in 1979. Two years later, he and the well-known securities analyst, Ben Rosen, pooled their resources to found Sevin-Rosen Partners, a high-tech venture-capital firm.
Robert J. Shillman, Ph.D. (Boston, Massachusetts)	54	His Cognex Corporation markets an optical character recognition system (DataMan) that reads and relays printed, embossed, and inscribed lettering.
Alan F. Shugart (Scotts Valley, California)	69	Shugart founded the disk drive company that still carries his name but was eventually forced out. His next co-venture, Seagate Technology, is the industry leader in the production of 5¼-inch Winchester disk drives. (See Finis F. Conner.)
Harold C. Simmons (Dallas, Texas)	69	His early years as a bank examiner and drugstore entrepreneur were but a prelude to his true vocation—as a corporate raider. When he takes a major position in a company's stock through his Contran Corporation, its management quakes with fear.
Melvin Simon (Indianapolis, Indiana; Beverly Hills, California)	73	Simon made his fortune building shopping centers. Now he's enamored with filmmaking and he's testing his creative mettle as an independent producer.
William (Bill) E. Simon (Morristown, New Jersey)	72	After a stint as U.S. Treasury Secretary, he's back to his first love—making money. He's building a diversified conglomerate (Wesray Corporation) by acquiring undervalued companies through leveraged buy-outs and debt financing. Oil brokerage and financial consulting are sideline businesses.

TABLE 9 *(Continued)*

Candidate (Principal Residence)	Age in Year 2000	Source or Sources of Wealth to Date
Frederick W. Smith (Memphis, Tennessee)	56	Smith used an inheritance of $3.2 million plus borrowed money to implement an idea he developed for a Yale economics course: to structure a nationwide overnight air courier service (Federal Express) so that all the planes make one daily trip to a central location to exchange cargo.
Merrill Solomon (Washington, D.C.)	56	Solomon was only twenty-four when he founded Penril Corporation, whose first product was an electronic funds-transfer system. Today he's an independent inventor.
Michael Jay Solomon (New York City, New York)	62	(See Michael N. Garin and Telepictures Corporation.)
Sheldon H. Solow (New York City, New York)	late 60s	Solow is assembling a fortune in Manhattan real estate.
James LeVoy Sorenson (Salt Lake City, Utah)	78	He claims he's the richest man in Utah. Sorenson's entrepreneurial credentials include Deseret Pharmaceuticals (sold out to his two partners); LeVoy's (lingerie and cosmetics manufacturing); and Sorenson Research (a medical equipment firm sold to Abbott Laboratories).
George Soros (New York City, New York)	70	Soros Fund Management owns several offshore mutual funds. Soros, a Hungarian immigrant with liberal ideas, channels some of his excess cash into a foundation aptly named The Open Society Fund.

Alex G. Spanos (Stockton, California)	76	His A. G. Spanos Construction was swept along by the California real-estate boom. With two sons and two sons-in-law in the business, Spanos finds time to swing a golf club these days and enjoy the view from his owner's box at San Diego Chargers football games.
Aaron Spelling (Bel-Aire, California)	72	Spelling is one of Hollywood's richest and most prolific independent purveyors of schlock TV programs.
Jerry Spiegel (Hicksville, New York)	70	He maintains a low profile even though he's the largest individual owner of commercial and industrial real estate on Long Island.
Steven Spielberg (Los Angeles, California)	53	Spielberg, a multitalented screenwriter/director/producer, is already superwealthy from his blockbuster hits *Jaws, Close Encounters of the Third Kind, Raiders of the Lost Ark, Poltergeist,* and *E.T.*
Charles Srebnik (New City, New York)	65	An investment banker by vocation, a cattle breeder by avocation, Srebnik founded Genetic Engineering, Inc. This Northglenn, Colorado-based R&D firm applies gene-splicing techniques to livestock breeding.
David Stamm (Sunnyvale, California)	47	(See Aryeh Finegold.)
Saul P. Steinberg (New York City, New York)	61	Steinberg started in the computer leasing business with Leasco but dreamed of taking over a fashionable property like Chemical Bank. When the banking establishment balked, he settled for Reliance Insurance, now the cornerstone of his private diversified empire.

TABLE 9 *(Continued)*

Candidate (Principal Residence)	Age in Year 2000	Source or Sources of Wealth to Date
Jackson T. Stephens (Little Rock, Arkansas)	76	He and an older brother own businesses ranging from investment banking to public utilities and insurance.
Barbra Streisand (Malibu Beach, California)	58	This Brooklyn-born entertainer has moved beyond mere performing. Now she produces movies and TV shows and is open to any promising new business venture.
Ken J. Susnjara (Dale, Indiana)	52	His Thermwood Corporation, originally a plastics producer, now sells robots to industrial equipment manufacturers that resell them under their own names.
Robert B. Sutton (Breaux Bridge, Louisiana)	67	After his personal bankruptcy in 1970, it's all been uphill. Today he's a prosperous oil reseller and producer.
Robert A. Swanson (South San Francisco, California)	53	(See Herbert W. Boyer.)
James M. Sweeney (Irvine, California)	57	(See Dick P. Allen.)
Stephen C. Swid (New York City, New York)	59	(See Marshall S. Cogan.)
Sirjang Lal (Jugi) Tandon (Chatsworth, California)	59	Jugi Tandon was born in India and educated in the United States. His Tandon Corporation makes drives for floppy as well as Winchester disks.

A. Alfred Taubman (Bloomfield Hills, Michigan)	75	Shopping centers form the basis of his swelling fortune. In 1977, he led the consortium that purchased the Irvine Ranch (Orange County, California) for development. An art collector by avocation, he jumped at the chance to play white knight in a Sotheby Parke Bernet take-over bid.
William J. Texido (San Francisco, California)	64	He left Itel to found BRAE Corporation in 1977. BRAE leases transportation equipment but is now diversifying.
William M. Theisen (Omaha, Nebraska)	54	Theisen started Godfather's Pizza, a franchised chain, in the early 1970s and expanded until it had restaurants in more than forty states. In 1983, his chain was gobbled up by an even larger one, Chart House, Inc.
R. David Thomas (Dublin, Ohio)	67	Wendy's International, the franchised hamburger chain, is Thomas's bid for fame and fortune.
James E. Thornton (Brooklyn Park, Minnesota)	75	With two partners, he founded Network Systems Corporation in 1974 to manufacture equipment to link dissimilar computers of all sizes in local area networks.
John M. Thornton (Chatsworth, California)	68	(See William A. Norred.)
William H. Tillson (San Francisco, California)	51	NETCOM International, his company, arranges for the satellite transmission of commercial, cable, and subscription TV broadcasts as well as closed-circuit videoconferences.
Laurence A. Tisch Preston Tisch (New York City, New York)	76 73	With their father's resort hotel as a base, the Tisch brothers built a lodging chain and then diversified into other businesses (theaters, tobacco, insurance).

TABLE 9 *(Continued)*

Candidate (Principal Residence)	Age in Year 2000	Source or Sources of Wealth to Date
Jack Tramiel (Santa Barbara, California)	72	The hard-driving, Polish-born Tramiel turned Commodore International into a feisty microcomputer firm that dominates the European market. In 1984 he resigned in a boardroom clash with the company's chairman. At issue: Tramiel's dictatorial management style. A few months later, he bought Atari, the video games and computer division of Warner Communications, for $240 million in long-term notes and warrants.
Henry Traub (Clifton, New Jersey)	73	Another Polish immigrant, Traub founded Automatic Data Processing, now one of the largest independent DP firms.
James G. Treybig (Cupertino, California)	59	With three partners, he founded Tandem Computers, the company that pioneered the concept of fail-safe minicomputer systems. The secret: built-in redundancy.
John E. Trombly (Burlington, Massachusetts)	52	(See Arthur L. Fox.)
Gerald Tsai, Jr. (New York City, New York)	71	Born in Shanghai, Tsai studied economics at Boston University, then became an outstanding money manager. Today this financial operator is the largest shareholder of and an executive with American Can Co.
Robert E. (Ted) Turner III (Atlanta, Georgia)	62	Using as a foundation his late father's billboard advertising company, Turner bought several TV stations. Later he formed Turner

		Broadcasting (he owns 87 percent equity) to offer cable TV network programming. His outgoing personality and straight-shooter talk make him a preeminent force in the industry.
Stephen C. Turner (Gaithersburg, Maryland)	55	Turner is the founder of Bethesda Research Laboratories, a biotechnology firm that sells products to other genetic engineering researchers as well as engaging in its own contract research.
Ralph K. Ungermann (Santa Clara, California)	58	Ungermann's first venture was Zilog, a semiconductor firm heavily backed by Exxon Enterprises. His present company, Ungermann-Bass, Inc., makes local area network systems. (See Charlie C. Bass.)
Donald T. Valentine (Menlo Park, California)	67	Through his venture-capital investments, he's a director and major stockholder in Pizza Time Theatre, Altos Computer, and LSI Logic Corporation, among others.
Jan Van Andel (Ada, Michigan)	76	(See Richard M. DeVos.)
Charlie Vaughn (Norcross, Georgia)	62	(See Don Hudgins.)
Manuel A. Villafana (St. Paul, Minnesota)	59	This medical-equipment tycoon sold his first company, Cardiac Pacemakers, to Eli Lilly, netting $3 million. His second, St. Jude Medical, makes artificial-heart valves.
Philippe Villers (Billerica, Massachusetts)	65	Villers, a Frenchman by birth, cofounded Computervision (CAD/CAM). When Computervision's management refused to diversify into robotics, Villers founded his own robotmaker, Automatix, Inc.

TABLE 9 *(Continued)*

Candidate (Principal Residence)	Age in Year 2000	Source or Sources of Wealth to Date
Alan Voorhees (Maynard, Massachusetts)	76	His Data Terminal Systems makes terminals and control systems.
Bruce W. Vorhauer, Ph.D. (Costa Mesa, California)	58	A biomedical engineer, Vorhauer founded VLI Corporation to market his over-the-counter contraceptive sponge, brand name Today. He also designed the equipment to manufacture the sponge.
David Wagman (Los Angeles, California)	48	(See Robert Leff.)
Cyril Wagner, Jr. (Midland, Texas)	66	(See Jack Brown.)
Harvey E. Wagner (Berkeley, California)	63	First came the founding of Teknekron (1968), a firm that invests in new R&D businesses. Next came Wagner's investment in Tera Corporation (1974). Tera sells an integrated line of packaged software aimed primarily at the electric utility industry. It also offers computer-aided engineering services.
Ben C. Wang (Culver City, California)	74	In 1969, Wang founded Wangco (peripheral computer equipment), purchased by Perkin-Elmer in 1976. Rexon, his current venture, manufactures microprocessor-based small-business computer systems as well as a family of ¼-inch streaming cartridge tape drives.

Albert S. Waxman, Ph.D. (Milpitas, California)	60	With backing from Arthur Rock and Robert Noyce, Waxman founded Diasonics in 1977. It specializes in computer-based ultrasound imaging systems as well as related equipment used to diagnose disease and organic disorders.
James Weersing (Mountain View, California)	60	This engineer with an M.B.A. from Stanford is following a unique get-rich strategy. To date he has founded three medical equipment companies (Sutter Biomedical, Sequoia Turner, and LifeScan), which he expects to sell eventually to large drug companies.
Sanford I. Weill (New York City, New York)	67	This financial operator has deftly maneuvered his way around the canyons of Wall Street, ending up as an executive and a major stockholder of American Express Co.
Walter K. Weisel (Kalamazoo, Michigan)	60	An entrepreneur in his own right (NC Services, Inc.), Weisel was enticed from Unimation and given substantial equity when he joined PRAB Robots as president.
Leslie H. Wexner (Columbus, Ohio)	63	His specialty clothing chain, The Limited, has over four hundred stores. Wexner bought Lane Bryant, an even larger chain.
John Whitney, Jr. (Los Angeles, California)	54	(See Gary Demos.)
Clayton Williams, Jr. (Midland, Texas)	68	This former insurance agent switched to the oil business in the late 1960s. He's made millions but he's also highly leveraged.
Kenneth Williams Roberta Williams (Coursegold, California)	46 47	From one good software package—a fantasy game she wrote, he programmed—doth a company grow. Sierra On-Line, Inc., is now one of the country's foremost video game software houses.

TABLE 9 *(Continued)*

Candidate (Principal Residence)	Age in Year 2000	Source or Sources of Wealth to Date
Warren Winger (Richardson, Texas)	57	Winger still owns nearly 30 percent of his publicly owned chain of company-owned retail computer stores, CompuShop.
Tadeusz Witkowicz (Worcester, Massachusetts)	50	(See Richard A. Cerny.)
Robert N. Wold (Los Angeles, California)	75	He lent his name to a company that brokers satellite transponder time. It also routes broadcast signals via terrestrial routes and will produce videoconferences for corporate clients.
Ervin Wolf	74	His Inexco Oil specializes in offshore drilling.
John Wolfe (Columbus, Ohio)	60s	Wolfe owns the *Columbus Dispatch.*
Stephen G. (the Woz) Wozniak (Berkeley, California)	50	The Woz designed the Apple I and II, got superrich when Apple Computer went public, retired from active management duty, and returned to college to get an undergraduate degree. Now he's back doing what he loves best—running a design team at Apple. (See Steven P. Jobs.)
Oscar Wyatt, Jr. (Houston and Corpus Christi, Texas)	76	He's the founder of Coastal States Gas Corporation.

Sam Wyly (Dallas, Texas)	66	Wyly is a venture capitalist who is heavily involved in the software and data processing industry.
Stephen Wynn (Las Vegas, Nevada)	58	Wynn is the protégé of E. Parry Thomas, an influential Nevada gaming financier. In 1973, Wynn took over the Golden Nugget casino, which remains his base of operations. His is the bankroll behind a software firm called Arktronics.
Barry Yampol (Englewood, New Jersey)	62	He founded Graphic Scanning Corporation in 1968 and built it into the largest radio-paging company in the country. He's diversified into other communications areas, including data transmission and cellular mobile radiotelephone service.
Alan (Bud) Yorkin (Los Angeles, California)	74	Yorkin, who got his start in TV as a director and writer, is Norman Lear's partner in Tandem Productions.
Ed Zaron (Baltimore, Maryland)	57	His MUSE Software specializes in educational and entertainment microcomputer programs.
Mortimer B. Zuckerman (Boston, Massachusetts)	62	Zuckerman, a lawyer by training, amassed a fortune in real estate. Then he turned altruistic as the financial savior of the venerable *Atlantic Monthly.*

Appendix

Survey Questionnaire

The following three-page questionnaire was mailed in mid-April 1982 to 2,000 forecasters. The 225 forecasters who filled in and returned the questionnaire fell into the specific categories of futurist (4 percent); venture capitalist (9 percent); securities analyst of selected high-growth industries (16 percent); academic who specializes in technological change (26 percent); business consultant engaged in strategic planning (15 percent); corporate long-range planner (17 percent); and government economist/trade publication editor (13 percent).

The survey results were compiled and analyzed by Richard Foy, Ph.D., and the research staff of Boyden Associates in New York City.

QUESTIONNAIRE

RAPID-GROWTH INDUSTRIES / COMPANIES

1. Today oil companies dominate the top 10 corporations on the *FORTUNE 500* list. What kind of corporation do you expect to dominate the top 10 in the year 2000?

2. What 5 technologies do you envision exerting the most economic influence over the next 20 years (i.e., 1980–2000)?

______________________ ______________________

______________________ ______________________

____

3. Check those industries or industry segments that will exhibit the highest profitability and/or growth over the next 20 years:

☐ AEROSPACE
- ____ airlines
- ____ satellites

☐ AUTOMATED EQUIPMENT
- ____ office
- ____ factory
- ____ consumer

☐ BROADCASTING
- ____ network
- ____ independent stations
- ____ cable tv broadcasting
- ____ cable tv programming

☐ COMPUTER HARDWARE
- ____ mainframes
- ____ minicomputers
- ____ personal computers
- ____ peripherals

☐ COMPUTER SOFTWARE
- ____ program creators/authors
- ____ software publishers
- ____ software retailing

☐ CONSTRUCTION/REAL ESTATE
- ____ building materials/equipment
- ____ home building
- ____ industrial construction
- ____ office construction
- ____ real estate

☐ CONSUMER ELECTRONICS
- ____ appliances
- ____ audio equipment (radios, stereos, etc.)
- ____ electronic games/toys
- ____ television
- ____ video recorder/playback equipment

☐ ELECTRONICS & COMPONENTS
- ____ micro computers
- ____ semiconductor memory
- ____ measurement & scientific equipment
- ____ tools

☐ ENERGY/MINING/PETROCHEMICALS
- ____ coal
- ____ chemicals/plastics
- ____ natural gas
- ____ nuclear
- ____ petroleum
- ____ solar
- ____ mining & drilling equipment

☐ FINANCIAL SERVICES
- ____ retail banking
- ____ brokerage
- ____ insurance

☐ FOOD/BEVERAGES
- ____ "acquaculture"
- ____ agriculture
- ____ commodities
- ____ distributors/wholesalers
- ____ processors
- ____ restaurants/fast food chains
- ____ supermarkets/retailers

☐ GENETIC ENGINEERING

☐ HEALTH CARE
- ____ medical equipment & supplies
- ____ medical services

☐ LEISURE/RECREATION
- ____ hotel/motel
- ____ gambling
- ____ recreational equipment & supplies
- ____ travel/vacation/resort

☐ MATERIALS
____ abrasives
____ cement
____ glass

☐ MOTION PICTURES
____ traditional film productions
____ film productions using advanced computer-based techniques (e.g., computer graphics)
____ distribution

☐ PHARMACEUTICALS
____ drugs manufactured using conventional production methods
____ drugs manufactured using bio-engineering production methods
____ retailing

☐ PUBLISHING/PRINTING
____ book
____ magazine & newspaper
____ electronic/data base

☐ TELECOMMUNICATIONS
____ data hardware
____ data transmission
____ voice hardware
____ voice transmission
____ satellite services
____ terrestrial services

☐ OTHER (describe): ____________________
____
____
____

4. List the 5 "fastest track" industries you checked in question #3 in descending order of their growth—the highest-growth industry or segment first. Next, answer the questions in each column:

RAPID-GROWTH INDUSTRY/ INDUSTRY SEGMENT	ESTIMATED % GROWTH FROM 1980–2000	IN THIS INDUSTRY, HOW FAST ARE TECHNOLOGICAL CHANGES OCCURRING? (very fast, moderately, slowly)	WHAT YEAR WILL THIS INDUSTRY MATURE? (i.e. reach a growth plateau & begin declining)	IS THIS INDUSTRY DOMINATED BY ENTREPRENEURIAL COMPANIES? (i.e. startups or small-to-medium-sized companies under 10 years old)
(1)				
(2)				
(3)				
(4)				
(5)				

5. Are any of the industries you listed in question #4 dominated by entrepreneurial start-ups or small-to-medium-sized companies under 10 years old? If so, name the industry's leading entrepreneurial companies and indicate each company's ownership status:

RAPID-GROWTH INDUSTRY DOMINATED BY ENTREPRENEURIAL COMPANIES	LEADING ENTREPRENEURIAL COMPANIES WITHIN THE INDUSTRY	THIS COMPANY IS: PUBLICLY-HELD CLOSELY-HELD PRIVATELY-HELD
(rapid-growth industry)		

(rapid-growth industry)		

(rapid-growth industry)		

ENTREPRENEURS

6. Identify any entrepreneurs in the rapid-growth industries of the future who have particularly interesting success stories I should highlight in my book:

Acknowledgments

I wish I could claim that the inspiration for *Future Rich* was mine alone. Alas, the credit belongs elsewhere.

The notion that my next book about the rich might offer a glimpse into the future rather than being a survey of the past was broached casually by Joseph Poindexter, a magazine editor with a keen ability to conceptualize innovative editorial ideas.

Although the proverbial light bulb immediately went on in my head, the idea frightened me as much as it intrigued me. Entrepreneurs were one thing, technical subjects quite another. How could I—a computer ignoramus not to mention a computer phobic—possibly tackle a book whose subtheme would be the high technology of tomorrow?

Another friend, Joseph Garber—a consultant by occupation, a futurist by inclination—gave me the courage to pursue the idea to its logical conclusion in the form of a book. Without his moral support and editorial suggestions this book would never have been written. Joe, a senior consultant for SRI International, read this book in every stage of its development, from the book proposal through the final manuscript. In particular, he checked all the technical descriptions throughout for accuracy. His many comments in the margins brought me back on course when I began veering off on tangents.

Meanwhile, I acquired a Xerox 820 microcomputer, which erased my phobia and speeded up the process of getting these words on paper. My machine also gave me an appreciation of the personal computer phenomenon. From firsthand experience I know why microcomputers are becoming household playthings, not to mention indispensable office productivity tools.

In this way, an offhand suggestion that at first struck me as totally impossible gradually evolved into one that seemed extremely difficult but doable and finally materialized as a challenge definitely worth the enormous research effort.

A number of other people helped me turn an untrained eye on the landscape of the future. Barbara Lowenstein, as much an editor as a literary agent, offered cogent comments that aided me in writing the sixty-page proposal that sold the publisher on the book's concept.

With the book contracted for, I then turned to other friends, who helped me with the grubby task of mailing out two thousand questionnaires to the country's small but elite fraternity of futurists. My mailing crew consisted of Walter and Denise Nervik, Chris and Ian Tornay, and Ted Hussa.

A debt of gratitude also goes to those 225 forecasters who took a half hour of their time to fill in the questionnaire, which several told me required as much hindsight as it did foresight; and to Dr. Richard Foy and the research staff at Boyden Associates, where the survey results were compiled and analyzed.

Because of its positive thrust, *Future Rich* is a book that enjoyed the cooperation of the majority of its subjects. After culling the names of outstanding young companies and entrepreneurs from the survey questionnaires, I wrote to the officers of each target company to garner more information. I was particularly interested in knowing the story of the company's founding; the founders' backgrounds; the company's principal investors; and the company's current ownership status.

The response to my request overwhelmed me. The material that poured into my office formed a stack six feet high and took me many weeks to skim and digest.

I am especially indebted to the many entrepreneurs who granted me personal and telephone interviews to flesh out their stories further.

For background information about companies and entrepreneurs who did not cooperate, I contacted the securities firm that had underwritten a company's initial public offering. All were extremely obliging, especially Eleanor Mascheroni of Prudential-Bache Securities' public-relations department. Prudential-Bache acted as the repository of last resort when I couldn't get a prospectus elsewhere.

My library research was made more pleasant by the gracious staff members at the General Society of Mechanics and Tradesmen Library in mid-Manhattan and the business division of the Brooklyn Public Library. In both places, my researcher, Cynthia Hussa, and I found the bulk of the articles and books that helped me penetrate the veil that hides tomorrow. In addition, I am obligated to the publicity directors at a number of major publishing houses. Upon request, they sent me just-published books on future-oriented and technological topics.

Trade shows and seminars were other sources of information for me. For permission to attend, I would like to thank the organizers of the following conferences: Information Utilities '82 in Rye, New York . . . Viewtext '82 in New York City . . . Videotex '82 in New York City . . . and Pratt Center's Computer Animation Seminar, also in New York City.

Several academics and government officials were particularly responsive when I approached them for information. For the light they shed on technical subjects, my appreciation goes to Professor Arnold C. Cooper of Purdue University's School of Management and Krannert Graduate School of Management; Richard W. Jensen, assistant chancellor at the University of California at Santa Barbara; John T. Scott, assistant professor of economics at Dartmouth College; David J. Ravenscraft, an economist on the staff of the Bureau of Economics at the Federal Trade Commission; Robert Ayres, professor of engineering and public policy at Carnegie-Mellon University; Daniel Peabody-Smidt of the Peabody Foundation for the Humanities; Dr. Harry M. Rosenberg, chief of the mortality statistics branch of the National Center for Health Statistics; and Francisco R. Bayo, deputy chief actuary at the Social Security Administration of the Department of Health and Human Services.

For explaining the intricacies of venture-capital financing, I owe a special thanks to Paul H. Stephens of the San Francisco investment banking firm of Robertson, Colman & Stephens. I would similarly like to thank Booz• Allen & Hamilton, Inc.'s Harvey Poppel, who promptly supplied me with various documents and studies on the information industry.

Closer to home, I want to express my obligation to my husband, William. He had the patience to listen to my ideas before they made their way to a piece of paper via my computer, and his layman's perspective on some rather technical subjects certainly helped me improve the final product.

Bibliography

[1] Adams, Russell B., Jr. *The Boston Money Tree.* New York: Thomas Y. Crowell, 1977.

[2] Allen, Frederick Lewis. *The Big Change: America Transforms Itself 1900–1950.* New York: Harper & Brothers, 1952.

[3] ———. *The Lords of Creation.* New York: Harper & Brothers, 1935.

[4] Asimov, Isaac. *Science Past, Science Future.* Garden City, N.Y.: Doubleday, 1975.

[5] Baer, Jean. *The Self-Chosen: "Our Crowd" Is Dead, Long Live Our Crowd.* New York: Arbor House, 1982.

[6] Beard, Charles A. and Mary R. *A Basic History of the United States.* New York: New Home Library, 1944.

[7] Bell, Daniel. *The Coming of the Post-Industrial Society.* New York: Basic Books, 1973.

[8] Bernstein, Jeremy. *The Analytical Engine: Computers—Past, Present and Future,* rev. ed. New York: William Morrow/Quill, 1981.

[9] Brandt, Steven C. *Entrepreneuring: The Ten Commandments for Building a Growth Company.* Reading, Mass.: Addison-Wesley, 1982.

[10] Brooks, John. *The Go-Go Years.* New York: Weybright and Talley, 1973.

[11] Brown, Ronald. *Telecommunications: The Booming Technology.* Garden City, N.Y.: Doubleday, 1970.

[12] *California's Technological Future: High Technology and the California Workforce in the 1980s.* Sacramento, Calif.: Department of Economic and Business Development, 1982.

[13] Cavalieri, Liebe F. *The Double-Edged Helix: Science in the Real World.* New York: Columbia University Press, 1981.

[14] Cetron, Marvin, and Thomas O'Toole. *Encounters with the Future: A Forecast of Life in the 21st Century.* New York: McGraw-Hill, 1982.

[15] Cherfas, Jeremy. *Man Made Life: An Overview of the Science, Technology and Commerce of Genetic Engineering.* New York: Pantheon, 1983.

[16] Clarke, Arthur C. *Profiles of the Future: An Inquiry into the Limits of the Possible,* rev. ed. New York: Harper & Row, 1973.

[17] Cochran, Thomas, and William Miller. *The Age of Enterprise.* New York: Macmillan, 1942.

[18] Connell, Stephen, and Ian A. Galbraith. *Electronic Mail: A Revolution in Business Communications.* New York: Van Nostrand Reinhold, 1983.

[19] Cornish, Edward, ed. *1999—The World of Tomorrow: Selections from THE FUTURIST.* Washington, D.C.: World Future Society, 1978.

[20] Culhane, John. *Special Effects in the Movies.* New York: Ballantine, 1981.

[21] Deken, John. *The Electronic Cottage: Everyday Living with Your Personal Computer in the 1980s.* New York: William Morrow, 1981.

[22] Engelberger, Joseph F. *Robotics in Practice.* New York: American Management Associations, 1980.

[23] Evans, Christopher. *The Making of the Micro.* New York: Van Nostrand Reinhold, 1981.

[24] ———. *The Micro Millennium.* New York: Washington Square Press/Pocket Books, 1981.

[25] Faulkner, Harold Underwood. *American Economic History,* rev. ed. New York: Harper & Brothers, 1931.

[26] Fishman, Katharine Davis. *The Computer Establishment.* New York: Harper & Row, 1981.

[27] Gardner, Ralph D. *Horatio Alger or the American Hero Era.* New York: Arco, 1978.

[28] Gilder, George. *Wealth and Poverty.* New York: Basic Books, 1981.

[29] Ginger, Ray. *The Age of Excess: The U.S. from 1877 to 1914.* New York: Macmillan, 1975.

[30] Goodman, George W. ("Adam Smith"). *Paper Money.* New York: Summit Books, 1981.

[31] Groner, Alex. *The American Heritage History of American Business and Industry.* New York: American Heritage Publishing, 1972.

[32] Hamilton, David. *Technology, Man and the Environment.* New York: Charles Scribner's, Sons, 1973.

[33] Hanson, Dirk. *The New Alchemists: Silicon Valley and the Microelectronics Revolution.* Boston: Little, Brown, 1982.

[34] Harsanyi, Dr. Zsolt, and Richard Hutton. *Genetic Prophecy: Beyond the Double Helix.* New York: Rawson, Wade, 1981.

[35] Hecht, Jeff, and Dick Teresi. *Laser: Supertool of the 1980s.* New York: Ticknor & Fields, 1982.

[36] Heyn, Ernest V. *A Century of Wonders: 100 Years of Popular Science.* Garden City, N.Y.: Doubleday, 1972.

[37] Holbrook, Stewart H. *The Age of the Moguls.* Garden City, N.Y.: Doubleday, 1953.

[38] Hutton, Richard. *Bio-Revolution: DNA and the Ethics of Man-Made Life.* New York: New American Library/Mentor, 1978.

[39] Josephson, Matthew. *The Money Lords: The Great Finance Capitalists 1925–1950.* New York: Weybright and Talley, 1972.

[40] ———. *The Robber Barons: The Great American Capitalists 1861–1901.* New York: Harcourt, Brace, 1934.

[41] Kidder, Tracy. *The Soul of a New Machine.* Boston: Little, Brown, 1981.

[42] Klein, Burton. *Dynamic Economics.* Cambridge, Mass.: Harvard University Press, 1977.

[43] Kotkin, Joel, and Paul Grabowicz. *California, Inc.* New York: Rawson, Wade, 1982.

[44] Krasnoff, Barbara. *Robots: Reel to Real.* New York: Arco, 1982.

[45] Lundberg, Ferdinand. *America's Sixty Families.* New York: Vanguard Press, 1937.

[46] ———. *The Rich and the Super-Rich.* New York: Lyle Stuart, 1968.

[47] Martin, James. *Telematic Society: A Challenge for Tomorrow.* Englewood Cliffs, N.J.: Prentice-Hall, 1981.

[48] Mensch, Gerhard. *Stalemate in Technology.* Cambridge, Mass.: Ballinger, 1979.

[49] Menshikov, S. *Millionaires and Managers.* Moscow, U.S.S.R.: Progress Publishers, 1969.

[50] Mills, C. Wright. *The Power Elite.* New York: Oxford University Press, 1956.

[51] Morison, Elting E. *From Know-How to Nowhere: The Development of American Technology.* New York: Basic Books, 1974.

[52] Myers, Gustavus. *History of the Great American Fortunes.* New York: Modern Library, 1936.

[53] Myers, Margaret G. *A Financial History of the United States.* New York: Columbia University Press, 1970.

[54] Naisbitt, John. *Megatrends: Ten New Directions Transforming Our Lives.* New York: Warner Books, 1982.

[55] Nelson, Theodor H. *Computer Lib.* Schooleys Mountain, N.J.: T. Nelson Publishing, 1974.

[56] Noble, David F. *America by Design: Science, Technology, and the Rise of Corporate Capitalism.* New York: Alfred A. Knopf, 1977.

[57] Nora, Simon, and Alain Minc. *The Computerization of Society.* Cambridge, Mass.: MIT Press, 1981.

[58] O'Connor, Richard. *The Oil Barons: Men of Greed and Grandeur.* Boston: Little, Brown, 1971.

[59] O'Neill, Gerard K. *2081: A Hopeful View of the Human Future.* New York: Simon & Schuster, 1981.

[60] Osborne, Adam. *Running Wild: The Next Industrial Revolution.* Berkeley, Calif.: Osborne/McGraw-Hill, 1979.

[61] Panati, Charles. *Breakthroughs.* New York: Berkley Books, 1981.

[62] Pursell, Carroll W., Jr., ed. *Technology in America: A History of Individuals and Ideas.* Cambridge, Mass.: MIT Press, 1981.

[63] Rheingold, Howard, and Howard Levine. *Talking Tech.* New York: William Morrow, 1982.

[64] Rodgers, William. *THINK.* New York: Stein and Day, 1969.

[65] Rogers, Michael. *Biohazard.* New York: Avon, 1979.

[66] Shook, Robert L. *The Entrepreneurs: Twelve Who Took Risks and Succeeded.* New York: Barnes & Noble Books, 1981.

[67] Shuman, James B., and David Rosenau. *The Kondratieff Wave.* New York: World Publishing, 1972.

[68] *The Smithsonian Book of Invention.* Washington, D.C.: Smithsonian Exposition Books, 1978.

[69] Sobel, Robert. *I-B-M: Colossus in Transition.* New York: Times Books, 1981.

[70] Stobaugh, Robert, and Daniel Yergin, eds. *Energy Future: Report of the Energy Project at the Harvard Business School.* New York: Ballantine, 1980.

[71] Thompson, Jacqueline. *The Very Rich Book: America's Supermillionaires and Their Money—Where They Got It, How They Spend It.* New York: William Morrow, 1981.

[72] Thorndike, Joseph J. *The Very Rich: A History of Wealth.* New York: American Heritage Publishing Co., 1976.

[73] *The Timetable of Technology.* New York: Hearst Books, 1982.

[74] Toffler, Alvin. *Future Shock.* New York: Bantam, 1971.

[75] ———. *The Third Wave.* New York: Bantam, 1981.

[76] Watson, James D. *The Double Helix: A Personal Account of the Discovery of the Structure of DNA.* New York: New American Library/Signet, 1968.

[77] Weil, Ulric. *Information Systems in the 80's: Products, Markets, and Vendors.* Englewood Cliffs, N.J.: Prentice-Hall, 1982.

Notes

The following abreviations, used throughout the Notes, stand for these sources:

AM	American Machinist	FTN	Fortune
BAR	Barron's	FW	Financial World
BW	Business Week	HBR	Harvard Business Review
CB	California Business		
CC	Creative Computing	HT	High Technology
CEN	Chemical & Engineering News	ICP	ICP Software Business Review
CGW	Computer Graphics World	IES	IEEE Spectrum
		INC	Inc.
CHW	Chemical Week	IW	InfoWorld
CM	California Magazine	LABJ	Los Angeles Business Journal
CW	Computerworld		
DBM	Dun's Business Month	LAT	Los Angeles Times
		M$	Money
DC	Data Communications	MMH	Modern Materials Handling
DT	Datamation		
EB	Electronic Business	MMS	Mini-Micro Systems
EET	Electronic Engineering Times	NB	Nation's Business
		NPR	National Productivity Review
EN	Electronic News		
ESQ	Esquire	NSWK	Newsweek
FBS	Forbes	NYT	The New York Times

PI	The Philadelphia Inquirer	SN	Science News
		TM	Time
RT	Robotics Today	TR	Technology Review
SA	Scientific American	USNWR	U.S. News & World Report
SC	Science		
SC82	Science '82	VTR	Venture
SJM	The San Jose Mercury	WP	The Washington Post
		WSJ	The Wall Street Journal
SMN	Smithsonian		

The bracketed [ ] numbers refer to books listed in the Bibliography. For example, when you see [2], it stands for the second item in the Bibliography, specifically: Allen, Frederick Lewis. *The Big Change: America Transforms Itself 1900–1950.* New York: Harper & Brothers, 1952.

Preface

Page

9 The term "postindustrial" was used first in the future-oriented literature in 1914. It appeared in the title of an anthology, *Essays in Post-Industrialism: A Symposium of Prophecy Concerning the Future of Society,* edited by Ananda Coomaraswamy and Arthur J. Penty (London: T. N. Foulis, 1914). More recently, Daniel Bell used the phrase in the title of his definitive 1973 book [7].

10 "The crucial point": Daniel Bell, "Communications Technology—For Better or for Worse," *HBR,* May–June 1979.

13 As a motivator: [27] and "Could Horatio Alger's Heroes Make It in Today's Business World?," Ralph D. Gardner (*TWA Ambassador,* January 1975).

13 As the sociologist C. Wright Mills pointed out: [50], pp. 102–7.

14 In a previous book: [71], pp. 123–28.

Introduction: Avenues to Wealth by the Year 2000

Page

28 For more detail about the long-wave hypothesis, consult [67]. Articles include: "Reindustrialization: Aiming for the Right Targets," Nathaniel J. Mass and Peter M. Senge (*TR,* August/September 1981); "Echoes from a Siberian Prison Camp," James W. Michaels, William Baldwin, and Lawrence Minard (*FBS,* November 9, 1981); and "The Passing of the Hydrocarbon Era," Craig S. Volland (*HT,* January 1983).

30 The contemporary German economist: [48]; and the article "Q&A with Gerhard Mensch, On Innovation in Bad Times" (*VTR,* September 1982).

31 "over the span of a century": [59], p. 36.

33 "It is continually changing": "America's Restructured Economy: Hungry for Capital to Sustain the Boom" (*BW,* June 1, 1981).

36 "a new sort of technology": [68], pp. 190 and 217.

37 Beginning . . . the Carter administration: "Future of Small Business May Be Brighter Than Portrayed," David E. Gumpert (*HBR,* July–August 1979); and "Urban Aid for Financing Small Businesses," Samuel S. Beard (*HBR,* November–December 1980).

37 Another alarm sounded: "Venture Capital Becoming More Widely Available," David E. Gumpert (*HBR,* January–February 1979).

38 Thus by 1983: "How High-Technology Ideas Often Become Reality in Silicon Valley," Carrie Dolan (*WSJ,* August 2, 1983).

38 But if the traditional sources: "Going Wild Over Going Public" (*BW,* December 6, 1982); "The Window's Open Again," Lori Ioannou (*VTR,* April 1983); and "Right Time for Going Public," Michael Blumstein (*NYT,* December 27, 1983).

38 "Save all the hard work": Speech made by Jerry K. Pearlman, senior officer of the Heath and Computer Business Group of Zenith Radio, at the 1981 Rosen Research Personal Computer Forum (May 11–13, 1981).

39 And Reagan let it be known: The conventional economic wisdom that in highly concentrated industries the leading companies reap above-average profits was borne out by studies done in the early 1950s, particularly those of University of California economist Joe S. Bain. But more recent studies argue that a company's market share is a primary determinant of the size of its profits, irrespective of how concentrated the industry.

In the spring of 1982, the Justice Department unveiled new merger guidelines taking this new thinking into account. It announced that it was discarding the old ambiguous method of evaluating industry concentration and replacing it with a new method that was strictly a matter of arithmetic. The traditional way to measure concentration had been to add the market shares of the four largest companies and then make an arbitrary decision about whether it was too concentrated. Henceforth, anyone could make a definitive judgment about whether an industry was too concentrated by consulting the Herfindahl Index. This index is derived by squaring the market share of each company in an industry and then adding up the totals. Under the new antitrust guidelines, any merger that leaves an industry with a Herfindahl Index of over 1000 is extremely likely to be challenged.

See "Attacking the Test That Curbs So Many Mergers" (*BW,* November 16, 1981) and "A Loosening of Merger Rules" (*BW,* May 17, 1982).

39 "long-term, high-risk research": Speech by Dr. George A. Keyworth II, science adviser to President Reagan, "3 Reagan Aides Affirm

Federal Support for Basic Scientific Research," Robert Reinhold (*NYT,* June 26, 1981).

39 Fortunately, American companies: "The Technology Race," William Stockton (*NYT Magazine,* June 28, 1981); "Spending for Research Still Outpaces Inflation" (*BW,* July 6, 1981); and "R&D Spending Surges" (*BW,* June 7, 1982).

40 In general, fledgling companies: For a discussion of large vs. small companies' relative advantages and disadvantages in exploiting the high-technology marketplace, consult "Small Companies Can Pioneer New Products," Arnold Cooper (*HBR,* September–October 1966); and *Technological Innovation: Its Environment and Management,* Robert Charpie and others (Washington, D.C.: U.S. Government Printing Office, 1967).

40 "Breakthrough research": "The Case for Technology Entrepreneurs," Henry E. Riggs (*The Stanford Engineer,* Spring/Summer 1980). Also see [42].

42 There are definite advantages: For insight into smaller companies' most effective market strategy in the face of potential competition from large corporations, see "Small Manufacturing Enterprises," W. Arnold Hosmer (*HBR,* November–December 1957); "Meeting the Competition of Giants," Alfred Gross (*HBR,* May–June 1967); and "In High-Tech Battles, Goliaths Usually Win by Outlasting Davids," Lawrence Ingrassia (*WSJ,* September 13, 1982).

44 The story was equally tragic: [60], pp. 23–25.

45 Surprising as it may be: "Lists" (*VTR,* December 1981), p. 63. For a discussion of the low failure rate among high-technology firms in the Boston area, see "Influences Upon Performance of New Technological Enterprises," Edward B. Roberts, in *Technical Entrepreneurship: A Symposium,* eds. Arnold Cooper and John Komives (Milwaukee: Center for Venture Management, 1972).

45 This is not a high failure rate: Dun & Bradstreet, *Patterns for Success in Managing a Business* (New York, 1967), p. 3.

45 "Although not represented": Albert V. Bruno and Arnold C. Cooper, "Patterns of Development and Acquisitions for Silicon Valley Startups," Reprint No. 867 (West Lafayette, Ind.: Institute for Research in the Behavioral, Economic, and Management Sciences, September 1982), p. 288.

46 "This guy had": Jerry Bowles, "Silicon Valley Days" (PMM&Co.'s *World 1982,* No. 2).

46 Today there are usually: "The Palo Alto Experience," Arnold C. Cooper (*Industrial Research,* May 1970); "Technical Entrepreneurship: What Do We Know?" Cooper (*R&D Management,* February 1973); and "Success Among High-Technology Firms," Cooper and Albert V. Bruno (*Business Horizons,* March 1977).

47 Many entrepreneurs are: Dave Lindorff, "No Limits to Growth" (*VTR,* May 1982).

47 "It's a waste": "The Thrill of Starting Up Again" (*BW,* May 4, 1981).

1 Biotechnology: The Breakthrough Wizards

Page

53 "I have a feeling": Harold M. Schmeck, Jr., "DNA's Code: 30 Years of Revolution" (*NYT,* April 4, 1983). For a firsthand account of the revolutionary discovery, read [76].

55 Throughout most of the 1970s: [38], [65], and [13].

56 But none of these obstacles: "Cetus Charting a Broad Course," Thomas J. Lueck (*NYT,* June 5, 1981); "Sit It Out or Sell Out," Thomas O'Donnell (*FBS,* March 1, 1982); "Cetus Reduces Staff and Cuts Programs" (*NYT,* September 8, 1982); and "New President of Cetus Is Products Specialist" (*NYT,* December 13, 1982).

58 Under the scrutiny: For more on Dr. Farley's resignation read "Founder of Cetus Starts New Venture" (*NYT,* May 10, 1983).

58 In 1975, Bob Swanson: "Genentech: Life Under a Microscope," Susan Benner (*INC,* May 1981). Also see: "The Gene Trust," Kathleen and Sharon McAuliffe (*Omni,* March 1980); "Here Comes the Test-Tube Shock Wave," David O. Weber (*CB,* October 1980); and "A Splice of Life," David Gilman (*World,* Spring 1981).

60 Other Genentech syntheses: Genentech listed the products it had in various stages of development in a 10-K report filed with the SEC (December 31, 1981), p. 2.

60 In the late 1970s: Two articles, in particular, showcase the greed of those seeking to profit from biotechnology—"Throngs Descend on Genetic Conference," Hal Lancaster (*WSJ,* May 4, 1981); and "Dog Days on Wall Street," Paul G. Brown (*FBS,* March 14, 1983).

61 Emblematic of Wall Street's: "Genentech Issue Trades Wildly," Tim Metz (*WSJ,* October 15, 1980); and "Genentech, New Issue, Up Sharply," Robert J. Cole (*NYT,* October 15, 1980).

61 Among the latter group: The tribulations of Southern Biotech are sketched in: "Southern Biotech Seeks Chapter 11 Protection" (*WSJ,* June 2, 1982) and "The Bio-Bubble" (*CM,* September 1982).

62 To that end, Genentech maintains: "Genentech Discusses Raising $50 Million by Partnership Plan" (*WSJ,* August 26, 1982).

63 The hullabaloo surrounding Humalin: Nicholas Wade, "Inventor of Hybridoma Technology Failed to File for Patent" (*SC,* Vol. 208, May 16, 1980).

63 But the need to pay: "An Academic Rush for a DNA Bonanza" (*BW,* December 21, 1981); "Patent for Products from Gene Splicing Is Delayed by U.S." (*WSJ,* July 2, 1982); "Stanford Loses Gene Patent Bid," Tamar Lewin (*NYT,* August 6, 1982); and "File Closed by Stanford on Gene-Splicing Patent," Tamar Lewin (*NYT,* December 9, 1982).

64 In addition, the U.S. Patent Office: See "The Patent Race in Gene-

Splicing," Tamar Lewin (*NYT,* August 29, 1982); and "The Fathers of Gene-Splicing Fight for Custody" (*BW,* January 10, 1983).

64 "The methods developed": Lewin, op. cit.

64 Litigating should be worth: "Licenses for Gene-Splicing," Barnaby J. Feder (*NYT,* August 19, 1981).

65 But Glick and Johnston also retained: Prospectus (September 29, 1982) for Genex's initial public offering; and the prospectus (April 15, 1983) for an offering of $1.5 million more shares.

65 Kevin Landry, a general partner: "When Investors Assemble Their Own Companies," Harrison L. Moore (*VTR,* October 1982).

66 "We can't wait": Kenneth B. Noble, "Venture Capitalists' New Role" (*NYT,* August 31, 1981).

67 The problem highlights: For a history of the link forged early in the twentieth century between university educators and corporations with heavy research requirements, read [56], pp. 128–47.

67 Over the years: The conflict between the universities that employ the country's premier molecular biologists and the bioengineering industry that rents the brains of those biologists for profit are outlined in: "Government Scrutinizes Link Between Genetics Industry and Universities," Robert Reinhold (*NYT,* June 16, 1981); "Business Rents a Lab Coat," David E. Sanger (*NYT,* February 21, 1982); "Corporate Links Worry Scholars," David E. Sanger (*NYT,* October 17, 1982); "Academic Research and Big Business: A Delicate Balance," Katherine Bouton (*NYT,* September 11, 1983); and "The Tempest Raging Over Profit-Minded Professors" (*BW,* November 7, 1983).

67 "The values which we see": Dr. Goldstein made these remarks during an episode ("Life: Patent Pending") of *Nova,* WGBH Educational Foundation's TV series.

69 Although some forty or fifty: "Harvard Drafts Rules on Conflicts of Interest" (*NYT,* October 4, 1981) and "Schools Study Ethics of Business-Aided Research" (*NYT,* April 8, 1983).

69 The financial payoff: Prospectus (March 22, 1983) for Biogen N.V.'s initial public offering.

70 In a novel arrangement: "Universities' Accord Called Research Aid," Ann Crittenden (*NYT,* September 12, 1981); and "Engenics Inc. Is Given $7.5 Million" (*WSJ,* September 15, 1981).

70 But monetary dreams: Wade, op. cit., p. 693.

71 Hybritech, bucking to capture: "How Kleiner Perkins Flies So High" (*BW,* January 24, 1983).

72 Hybritech quickly moved out: "Medicine's New Pioneers," Paul Franson (*VTR,* August 1979); "Small Firm's Biotech Coup Wins Approval," (*INC,* January 1982); a Robertson Colman Stephens institutional research report (February 4, 1982); and the prospectus (October 28, 1981) for the company's initial public offering.

73 David Padwa is an unlikely: "Agriculture's Technology Revolution," David Weber (*VTR,* August 1981).

74 "After all," he says: " 'Nine Meals Away from Murder,' " Lisa Gross (*FBS,* March 1, 1982).

2 Hardware Horizons: The Computer Manufacturers

Page

77 Meanwhile, at the peak: Among the sources for the Apple Computer story are company-issued literature as well as articles and stock analysts' reports. Two securities analysts' assessments of Apple Computer are especially insightful: Esther Dyson's report No. 82-319 on Apple Computer (New York: Oppenheimer & Co., April 30, 1982), and the section on Apple Computer (pp. 160–72) in [77].

79 "friendly" microcomputer: In one of the earliest "whole earth catalog"-type books [55] aimed at the hobbyist underground, Theodor Nelson refers to any computer system that is "clear, easy to use, and friendly" as a "good-guy system." His glossary of "The Most Important Computer Terms for the '70s" lists other phrases synonymous with "good-guy system"—"user-oriented," "user-level system," "naïve-user systems," and "idiot-proof system." Clearly, the term "user-friendly" did not enter common computer parlance until later in the decade. See [55], p. 13.

82 In 1976, a year after: Telephone interview with Edward Faber (February 18, 1983). Also, "Store a Day" (*FTN,* September 19, 1983); and "The Instant Billionaire," Kathleen K. Wiegner (*FBS,* December 5, 1983).

85 There were other similarities: The facts about Henry Ford and the Ford Motor Company came from [2], pp. 7, 109–14, 124–25, and 130; and [62], pp. 163–75.

85 "Just aiming for": Sklarewitz, op. cit.

85 In another interview: Allan Tommervik, "Exec Apple: New President Mike Markkula" (*Softalk,* July 1981).

86 Again, times have changed: Apple Computer prospectus for first public offering of 4.6 million shares (December 12, 1980). Also, the prospectus (May 28, 1981) for a secondary offering of 2.6 million shares.

89 One commentator: Everett T. Meserve, "A History of Rabbits" (*DT,* September 1981).

90 The exception was: The story of IBM and the Watson family is documented in [64] and [69]. In addition, see "THIMK: Is Something Wrong at IBM?" (*FBS,* September 1, 1971).

90 Chicago-born Max Palevsky: [26], pp. 217–225; "Chapter Two" (*FTN,* March 12, 1979); and "Palevsky Rides Again," Maurice Barnfather (*FBS,* November 24, 1980).

92 But for Arthur Rock: Michael Moritz, "Arthur Rock—'The Best Long-Ball Hitter Around' " (*TM,* January 23, 1984).

92 William C. Norris, another would-be millionaire: [26], pp. 45–46 and 199–209. Also see "No Limits to Growth," Dave Lindorff (*VTR,* May 1982); "Seeking to Aid Society, Control Data Takes on Many Novel Ventures," Lawrence Ingrassia (*WSJ,* December 22, 1982); and "Control Data—Is There Room for Change After Bill Norris?" (*BW,* October 17, 1983).

93 Success invariably spawns imitators: [26], pp. 207–208; [77], pp. 129–36; and "Designer of Big New 'Super Computer' Prepares to Take His Company Public," Scott R. Schmedel (*WSJ,* February 2, 1976). For an explanation of the Control Data-Cray rivalry, read "Control Data and Cray Start Feuding Again," Lawrence Ingrassia (*WSJ,* May 25, 1982).

94 "IBM is wonderful": Personal interview (July 7, 1982).

94 His words have an ironic ring: "Trilogy Ltd. Delays Its New Computers" (*WSJ,* May 14, 1984); "Mainframe Plan Ended by Trilogy," David E. Sanger (*NYT,* June 12, 1984); and "Can Troubled Trilogy Fulfill Its Dream?" Eric N. Berg (*NYT,* July 8, 1984).

94 Dr. Gene M. Amdahl: "Taking on the Industry Giant" (*HBR,* March–April 1980); "The Second Time Around" (*FBS,* September 15, 1980); "Keeping Control" (*FTN,* September 22, 1980); "Gene Amdahl Programs Bold Goals into His Infant Computer Enterprise," Marilyn Chase (*WSJ,* October 9, 1980); and "Entrepreneurs Share Secret for Success," Leslie Conley (American Electronics Association newsletter *Update,* May 1982).

94 Gene Amdahl still claims: The details of Trilogy's financing are outlined in a comprehensive proposal booklet issued by the company in January 1982. Trilogy Computer Development Partners Ltd., a limited R&D partnership, is described in a Merrill Lynch White Weld Capital Markets Group prospectus (August 6, 1981) and in "Gene Amdahl's New Computer Venture," Robert J. Cole (*NYT,* August 23, 1981). Also see the prospectus (November 9, 1983) for an initial public offering.

96 Despite his son Carlton's youth: Michelle Bekey, "Born and Bred Entrepreneurs" (*VTR,* March 1981).

98 Tandem Computers, Inc.: "Tandem Has a Fail-Safe Plan for Growth," Susan Benner (*INC,* June 1981); "Tandem Computers—No Recession Here," Michael S. Malone (*NYT,* April 25, 1982); "Beyond the Better Mousetrap," Kathleen K. Wiegner (*FBS,* June 22, 1981); "Managing by Mystique at Tandem Computers," Myron Magnet (*FTN,* June 28, 1982); and [77], pp. 139–43.

99 Stratus Computer: The birth of Stratus Computer, Inc., is described in a Harvard Business School case study, No. 682–030 (1981); "A Fail-Safe Entry That's a Bargain" (*BW,* November 16, 1981); and "Stratus's Nonstop Computers," N. R. Kleinfield (*NYT,* July 11, 1984). Also see the prospectus (August 1983) for an initial public offering.

100 Synapse Computer Corp.: "Starting on a Fast Track," Jon Levine (*VTR,* November 1981); "Multi-68000 Transaction CPU Sports 'Additive' Architecture," Larry Lettieri (*MMS,* September 1982); and "The New Face of Manufacturing," Kevin Farrell (*VTR,* October 1982).
101 "about 30 megabucks": Speech, "The Personal Computer: A New Medium," given at the Rosen Research Personal Computer Forum (May 11–13, 1981).
102 Dr. Adam Osborne: "Big Plans for Little Computer" (*NYT,* February 15, 1982); "Osborne: From Brags to Riches" (*BW,* February 22, 1982); "A Computer Gadfly's Triumph," Bro Uttal (*FTN,* March 8, 1982); "Interview with Adam Osborne," Jack B. Rochester (*CW,* November 17, 1982); and "Other Maestros of the Micro," Frederic Golden (*TM,* January 3, 1983).
103 At the zenith: "Snags in Introducing New Computer Sidetrack Osborne's Stock Offering," Erik Larson (*WSJ,* July 13, 1983); "Pacesetting Computer Maker Files for Bankruptcy," Eric N. Berg (*NYT,* September 15, 1983); and "Behind the Big Collapse at Osborne," Fred R. Bleakley (*NYT,* November 6, 1983).
104 "You've got to fail": Erik Larson, "Osborne Takes Little of the Blame" (*WSJ,* October 13, 1983).

3 Computer Parts and Peripherals: The Suppliers

Page
106 "We are the salvation": "Can U.S. Chip Makers Afford the Industry's New Look?" (*BW,* January 18, 1982).
106 As Intel's production czar: [33], p. 125.
106 Finding buyers for: [33], pp. 115–27; "The Microprocessor Champ Gambles on Another Leap Forward" (*BW,* April 14, 1980); "For Intel, a Vote of Confidence," Thomas J. Lueck (*NYT,* December 23, 1982); "Problem-Plagued Intel Bets on New Products, IBM's Financial Help," Marilyn Chase (*WSJ,* February 4, 1983); and "Why They're Jumping Ship at Intel" (*BW,* February 14, 1983).
107 with the 8008: The 8008 was upgraded, or redesigned, several times to eliminate architectural boondoggles. The upgraded versions were given the numbers 8080 and 8080A, successively. See [60], pp. 25, 29, and 163.
108 At times, Intel's genius: [60], pp. 164–65.
109 The micromainframe is one: Gene Bylinsky, "Intel's Biggest Shrinking Job Yet" (*FTN,* May 3, 1983).
111 "Besides," said W. J. Sanders: "Why the Time Is Right Again for Semiconductor Ventures" (*BW,* January 18, 1982).
111 In the early 1980s: Geoff Lewis, "Why No One Wins Trade Secret Suits" (*VTR,* January 1983).

112 Jerry Sanders: "The Metamorphosis of a Salesman," Peter J. Schuyten (*NYT,* February 25, 1979); "Salesman Rampant, on a Field of Chips," Kathleen K. Wiegner (*FBS,* December 8, 1980); and "AMD Swaggers Toward 'Gigabuck' Proportions" (*BW,* February 21, 1983).

115 Before they'd even made: "Bucking the Odds in Standard Chips," Jon Levine (*VTR,* March 1982); and the profile of SEEQ in the newsletter, "Semiconductor Industry and Business Survey" (Sunnyvale: HTE Management Resources, August 30, 1982).

115 "We think there's": Andrew Pollack, "Singing the Semiconductor Blues" (*NYT,* May 24, 1981).

115 The legal skirmishing: Jon Levine, "Trade Secret Suit: Intel, SEEQ Settle Out of Court" (*VTR,* April 1982).

116 Rather than hinder SEEQ: "Investing in the West," Timothy Gartner (*CB,* February 1982); and a speech, "Trends in Non-Volatile Technology," given by SEEQ's President Gordon A. Campbell at the Rosen Research Seminar (May 15, 1982).

116 But no matter: See prospectus (October 12, 1983) for SEEQ's initial public offering.

117 Most people in: "Big Memories for Little Computers," Peter Nulty (*FTN,* February 8, 1982).

117 Winchester memory devices: There's some dispute about the origin of the name "Winchester." Some say it was the name of the street where IBM's laboratory was located. Others say the name is borrowed from the famous Winchester 30-30 rifle, an allusion to the fact that the earliest Winchester prototype contained two drives that could each store thirty megabytes of data (about thirty million characters or numerals) on rigid metal disks.

120 When Seagate went public: The prospectus (August 3, 1982) for a secondary offering of Seagate common stock.

120 C. Norman Dion: Roger Neal, "Millionaire Maker" (*FBS,* September 12, 1983).

121 "We had always": A speech given by Alan Shugart to the Kenna Club (April 1982).

121 Tandon Corporation was founded: Tandon's 1981 annual report; "New from Silicon Valley—The Winchester Drive," Michael S. Malone (*NYT,* January 10, 1982); "Tandon Corp.—Computers on Hot Roll and Parts Pay Off," Louis Sahagun (*LAT,* January 18, 1982); "Tandon Rides Computer Crest," Thomas C. Hayes (*NYT,* July 25, 1983); and "The Hard Driver Atop the Disk-Drive Heap," Kathleen K. Wiegner (*FBS,* November 7, 1983).

122 "Before the visit": Cathy Louise Taylor, "How Jugi Tandon Drives His Computer Firm to Record Sales Highs" (*LABJ,* November 9, 1981).

123 Tandon stock certificates: Prospectus for a November 24, 1981, issue of 1.8 million shares of Tandon common stock.

123 The disk-drive manufacturing business: "Keeping STC in the Fast Lane" (*BW,* June 7, 1982); and "Hope at Storage Technology," Thomas C. Hayes (*NYT,* March 21, 1983).

124 He'd long felt: "Storage Technology and Du Pont to Make Laser Systems for Data" (*WSJ,* June 21, 1983); and "Storage Technology Says Its New Gear Will Best IBM," G. Christian Hill (*WSJ,* September 22, 1983).

124 But apparently even Louisville: "Storage Technology Has Settled Lawsuit Against Ibis Systems" (*WSJ,* July 9, 1982).

126 Printronix, Inc.: See the prospectuses (June 17, 1980 and May 24, 1982) for two offerings of Printronix common.

126 David Lee: "At Qume, A Case in Point" (*NYT,* August 17, 1980); "Daisy Wheel Printer—Race for Sales" (*NYT,* October 29, 1981); "How to Make It Big—Engineers as Entrepreneurs," Tekia S. Perry (*IES,* July 1982); and the profile of Schroeder contained in "How the Top Students at Harvard Business Fare 20 Years Later," Earl C. Gottschalk, Jr. (*WSJ,* December 20, 1982).

4 Software: Its Publishers and Servicers

Page

129 Personal Software, Inc.: The profile of Daniel Fylstra and VisiCorp is based on an extensive personal interview with Fylstra (July 8, 1982), company handouts, and a variety of published sources.

134 During the heyday: "Software: New Territory for Venture Capitalists" (*BW,* October 19, 1981); "5,000 Entrepreneurs . . . and Counting," Jon Levine (*VTR,* January 1982); "New Fund Focuses on Software Firms" (*VTR,* February 1982); and "Software Companies Attracting Investors," William M. Bulkeley (*WSJ,* January 3, 1984).

135 ". . . Let's just say": Personal interview (July 8, 1982).

138 Mitch Kapor: "How Programmers Get Rich," Alexander L. Taylor III (*TM,* December 13, 1982).

140 By mid-1983, 1-2-3: Prospectus (October 6, 1983) for Lotus Development Corporation's initial public offering; "Venture Capital Partner Is Doing Fine, Thanks" (*NYT,* October 27, 1983); and "The Next Big Test for Lotus," David E. Sanger (*NYT,* February 13, 1984).

140 Strange as it may seem: For a brief description of the evolution of the stored program, consult [24], pp. 38–41; [23], pp. 83–87; and [8], pp. 70–74.

142 By 1980, the association: "The Top 20 in Computer Services" (*FBS,* July 6, 1981).

142 One of the very first: "ADP Trades Up," Harold Seneker (*FBS,* July 6, 1981); "How to Manage a Revolution," Martin Mayer (*FBS,* April 23, 1984); and three books—[26], pp. 280–87; [5], pp. 78, 80, 90–92, 95–96, 99, and 299–300; and [77], pp. 66–71 and 194.

143 As IBM goes: IBM's motivation for, and the effect of, "unbundling" as well as its forced spin-off of its DP service bureau are described in [26], pp. 135–38, 205–7, and 283.

143 Management Science America, Inc.: "The Leader's Drive to Stay Out Front" (*BW,* November 30, 1981) and "Software Superstar" (*M$,* February 1983).

144 MSA's strategy duplicates: The merger mania gripping the software industry is described in "When Acquisition Is Inevitable," Joanmarie Kalter (*VTR,* October 1981); "Why Programmers Buy Market Share," John Verity (*VTR,* December 1982); and "Slugging It Out on the Software Front," Andrew Pollack (*NYT,* October 16, 1983).

145 Over the past decade: MSA's prospectus (May 25, 1982) for a common stock offering.

145 Cullinet Software, Inc.: [77], pp. 32–43; " 'This Ought to Be a Business,' " Harold Seneker (*FBS,* July 6, 1981); "Masterminding a Software Empire" (*BW,* April 18, 1983); "Cullinet Offers Computer Link" (*NYT,* April 20, 1983); and "Integrated Data Base Link Ties Micros, Mainframes," Paul Gillin (*CW,* April 25, 1983).

146 When Cullinet floated: See the prospectus (August 2, 1978) for the firm's initial public offering.

146 But back in the late 1960s: "The Jim and Frank Show," Willie Schatz (*DT,* February 1982); "McCormack & Dodge: Innovating to Survive" (*ICP,* Autumn 1982); and "Microcomputers Gaining Primacy Forcing Changes in the Industry," William M. Bulkeley (*WSJ,* January 13, 1983).

146 Sandra Kurtzig: Personal interview with Ms. Kurtzig (July 6, 1982). Also see "A High Demand for Sandra Kurtzig's Software Prompted Her to Found ASK Computer," Joan A. Tharp (*The Executive,* July 1981); and "She Programs a Success Story," Andrew Pollack (*NYT,* October 13, 1981).

147 Kurtzig has managed: Prospectus for initial public offering (October 1, 1981).

148 "It's something": "Ask, and Ye Shall Receive" (*SJM,* February 22, 1982).

149 "I keep thinking": Personal interview (July 6, 1982).

151 CP/M's author, Dr. Gary Kildall: Digital Research corporate backgrounder (June 1982); "Venture Capital Firms Adopt Trailblazing Role" (*NYT,* August 1981); "Digital Resources Driven by Success into Major Changes" (*NYT,* September 15, 1981); "Digital Research and CP/M Move into the Big Time," Larry Lettieri (*MMS,* November 1981); and "Hit Programs: A Preview of Some Promising Software Companies," Monteith M. Illingworth (*BAR,* January 31, 1983).

153 Microsoft Corporation: Material supplied by the company as well as "A Software Whiz Kid Goes Retail" (*BW,* May 9, 1983); "A Leader in Computer Software, Microsoft Pushes . . . ," Erik Larson (*WSJ,*

December 23, 1983; and "Microsoft's Drive to Dominate Software," Stratford P. Sherman (*FTN,* January 23, 1984).

5 Conveying a Zillion Conversations: Ma Bell and Competitors

Page

159 Of the many ironies: The AT&T stock certificate is described in "The Bluest of the Blue Chips" (*TM,* January 25, 1982).

159 How rich was Bell?: Alexander Graham Bell's lack of long-term personal financial vision is described in [62], pp. 110–12.

161 Today the semiretired Carter: Andrew Pollack, "The Man Who Beat A.T.&T." (*NYT,* July 14, 1982). A full description of AT&T's trials and tribulations, including its brush with Carter, is contained in [26], pp. 291–320.

163 "Their whole hope": John Greenya, "The Revenge of the Long-Distance Inventor" (*Regardie's,* January–February 1981). The description of Goeken's involvement with MCI is also based on a telephone interview with him (March 2, 1983) and the article "The Un-Telephone Company" (*Communications,* September 1969).

164 Hailing from the same: [66], pp. 139–56; "The Man Who Beat Ma Bell" (*FTN,* July 14, 1980); "Battling Big AT&T, Little MCI Keeps on Landing Sharp Blows," Bernard Wysocki (*WSJ,* September 28, 1981); "Getting Stronger All the Time" (*FW,* May 1, 1982); " 'Find Me $16 Billion,' " Barbara Ettorre (*FBS,* June 7, 1982); "More Than Cheap Talk Propels MCI," Brian O'Reilly (*FTN,* January 24, 1983); and "MCI Races the Clock . . .," Virginia Inman (*WSJ,* June 14, 1983).

165 "I was really enjoying": [66], p. 143.

169 For MCI, the 1978 decision: "A New Breed of Phone Firm Starts to Grow," Jody Long (*WSJ,* April 29, 1982); "A Fleeting Market for AT&T's Resellers," G. Thomas Gibson (*VTR,* October 1982); "Bell's Long-Distance Battle," Andrew Pollack (*NYT,* October 13, 1982); and "Riding Consumer Coattails," Pamela Sherrid (*FBS,* January 30, 1984).

171 Take New York City: John Greenwald, "New Bells Are Ringing" (*TM,* October 25, 1982).

173 "I've been in government": John W. Dizard, "Gold Rush at the FCC" (*FTN,* July 12, 1982).

173 When the dust settled: "Discord Over Mobile Phones," Ernest Holsendolph (*NYT,* December 1, 1981); "The Looming Boom in Beepers and Car Phones," Aimee L. Morner (*FTN,* December 14, 1981); "Mobile Phone License Battle Picks Up Steam," Margaret Garrard Warner and James A. White (*WSJ,* June 7, 1982); and "Cellular-

Radio License Plan Proceeding . . .," Robert E. Taylor (*WSJ,* February 7, 1983).

174 "My business is": Andrew Pollack, "Graphic Scanning's Quiet Rise" (*NYT,* September 8, 1981).

174 Unfortunately for Yampol: "Graphic Scanning Is Investigated by FCC Over Radio Paging License Applications" (*WSJ,* August 25, 1982).

174 While the RCCs: The consortium for radiotelephone licenses formed by AT&T and the other WCCs is described in "AT&T Outlines Plans for New Subsidiary . . ." (*WSJ,* May 26, 1982); "Cellular Phone Service Plan for 30 Cities Will Give AT&T, GTE Uncontested Bids," Margaret Garrard Warner (*WSJ,* June 9, 1982); "Bell-GTE Cellular Radio Plan Assailed," Ernest Holsendolph (*NYT,* June 10, 1982); "GTE's Cellular Pact with AT&T" (*NYT,* June 22, 1982); and "AT&T Will Provide Mobile Phones in City," Ernest Holsendolph (*NYT,* February 19, 1983).

174 In October 1983: "FCC Licenses First Cellular Radio System," Jeanne Saddler (*WSJ,* October 7, 1983); and "Cellular Mobile Phone Debut" (*NYT,* October 14, 1983).

175 Apparently the FCC agrees: "Paging Frequencies" (*NYT,* April 30, 1982).

175 As usual, the companies: "Radio-Pager Market Booms as New Gear, Long-Distance Beeping Improve Service," Raymond A. Joseph (*WSJ,* November 30, 1981); "Lowly Beeper Coming of Age," Andrew Pollack (*NYT,* July 15, 1982); "Radio Pagers Expand Horizons," John G. Posa (*HT,* March 1983); and "The Beep Generation," Bernice Kanner (*New York* magazine, August 15, 1983).

176 Bill McGowan, too: "MCI to Buy Xerox's WUI . . ." (*WSJ,* December 16, 1981); "MCI Planning Outlay of $130 Million . . ." (*WSJ,* June 16, 1982); "MCI Is Cleared by FCC to Buy Xerox's WUI" (*WSJ,* June 24, 1982); "4 Establish Company for Paging" (*NYT,* July 29, 1982); and "MCI Aims at Market for Pagers," Philip H. Dougherty (*NYT,* October 19, 1982).

176 David Post of PageAmerica: "RCA and PageAmerica Plan a Joint Venture for Message Services" (*WSJ,* May 16, 1983); and "PageAmerica and RCA Globcom Announce the World's First International Personal Communications Network," Stuart Crump, Jr. (*Personal Communications,* July–August 1983).

176 Jack Goeken: "Airlines May Soon Be Offering In-Flight Telephone Service" (*WSJ,* July 31, 1981); "Communications: A Telephone for Airplanes," Andrew L. Yarrow (*NYT,* May 2, 1982); "Flying Phones," Letters to the Editor column (*NYT,* May 9, 1982); "Telephone Fly High," Herb Brody (*HT,* September/October 1982); and "Airfone Wires the Friendly Skies" (*VTR,* August 1984).

6 Switching On: The Telequipment Tycoons

Page

178 Cohen and two partners: "His 'Better' Phone Idea Grew into a Big Business," John Briggs (*Camden Courier-Post,* March 1, 1976); "Ma Bell Doesn't Faze TeleSciences," Mario A. Milletti (*Philadelphia Evening Bulletin,* June 9, 1977); "TeleSciences Rings Right Number with Growing Line-Up of Products," Michael Bassett (*BAR,* August 15, 1977); "A Small Firm Finds the Line to Big Profits," Dominic Sama (*PI,* February 20, 1979); "He Found a Better Way," Barry Crickmer (*NB,* August 1981); and "Millions Spent on Research Now Paying Off," James A. Walsh (*Camden Courier-Post,* April 18, 1982).

179 Because AT&T was still: "A Reluctant David Beats A.T.&T.," Andrew Pollack (*NYT,* December 13, 1981); "Reluctant David Bests Bell," Todd Mason (*Fort Lauderdale News/Sun-Sentinel,* January 3, 1982).

180 Second, with about 15 percent: "ROLM: By Any Other Name," Nancy Daly (*Sallyport,* Rice University magazine, November 1978); "A Hot New Challenger Takes on Ma Bell" (*BW,* February 12, 1979); "Silicon Valley Success Story" (*NYT,* June 4, 1979); "Aiming High from the Start, ROLM Already Has Sights Set on Billion-Dollar Mark," Susan A. Thomas (*San Francisco Business Journal,* July 7, 1980); "For ROLM, Success as Number 2," Merrill Brown (*WP,* June 14, 1981); "M. Kenneth Oshman Tackles Telecommunications Goliaths," Joan A. Tharp (*The Executive,* March 1982); and "Japan PBX Market Cracked by ROLM," Steve Lohr (*NYT,* January 16, 1984).

183 "We already had eighty percent": Adrian B. Ryans, Graduate School of Business case study of ROLM Corp. (Palo Alto: Stanford University, 1979), p. 3.

183 Still, the founders: The PBX manufacturers' bid to become the nerve center of the electronic office is described in "PBX: Emerging Hub for Electronic Offices," Dale Zeskind (*HT,* March 1983); "Why PBX Makers Are Sprouting So Fast" (*BW,* April 18, 1983); and "PBXs—at the Eye of the Phone Net Hurricane" (*HT,* May 1983).

184 ROLM's CBX II: "ROLM Introduces Data-Switching System . . . ," Carrie Dolan (*WSJ,* November 18, 1983); and "An Advanced PBX Shows Why IBM Bought into ROLM" (*BW,* November 28, 1983).

185 Alexander Graham Bell's photophone: Forrest M. Mims III, "The First Century of Lightwave Communications" (*International Fiber Optics and Communications,* February 1982).

185 In addition to being: The history of developing fiber optics technology is described in Mims, op. cit.; [35], pp. 82–86; and "Letters to *Fortune*" (*FTN,* May 17, 1982).

187 one Merrill Lynch study predicts: Glenn R. Pafumi, "Telecommunications/Technology: A Merrill Lynch Industry Study" (New York: Merrill Lynch, May 1981), pp. 23–54.
187 The first small-scale use: Les Brown, "TV's Use of Fiber Transmission Begins" (*NYT,* July 9, 1976).
187 The opportunities in fiber optics: "Fiber Optics: The Boom Comes Sooner" (*BW,* May 18, 1981); and "Revolutionizing Communications," G. Thomas Gibson (*VTR,* October 1981).
187 In Signal Hill: A long letter from Ted Naugler (September 22, 1982) outlined AMFOX's genesis, goals, and ownership.
188 Also based in the Los Angeles area: The source for the information on Lightwave was a September 17, 1982, letter from Dr. Dabby in which he included a biographical sketch. Also see the preliminary prospectus, dated June 27, 1984, for Lightwave Technologies Inc.'s initial public offering.
188 Richard Cerny: The mention of Artel is based on company product literature and press releases and the article "Keeping an Eye on the Future," Eileen Brill (*Business Worcester,* April 1982).
188 Not far away: The SpecTran mention is based on company literature and the prospectus for the initial public offering in May 1983.
188 Norrsken Corporation: The Norrsken sketch is based on two letters from Dr. Randall (September 23, 1982, and November 1, 1982).
189 And then there's: Math Associates was profiled in "He Sees the Light with Fiber Optics," Lawrence Van Gelder (*NYT,* March 2, 1980); and "Irwin Math of Math Associates," Richard L. Molay (*EET,* February 1, 1982).

7 Moving Pictures: The Electronic Entertainers

Page
194 "The joke in the electronics industry": Aljean Harmetz, "Home Video Games Nearing Profitability of the Film Business" (*NYT,* October 4, 1982).
194 Apparently the American military: "Video Games Go Marching Off to War," Francis X. Clines (*NYT,* July 25, 1982); and "Army Trying Out Video War Games" (*NYT,* November 20, 1983).
194 "It's important to have training devices": "Invasion of the Video Creatures" (*NSWK,* November 16, 1981).
195 Thus it should come: The creation of Simon, Merlin, and the lesser electronic toys emanating from Milton Bradley and Parker Brothers, two New England companies whose rivalry dates back to the nineteenth century, is covered in Diane McWhorter's article "Electronic Shock in Toyland" (*Boston Magazine,* October 1978).
196 The Doyles are using this grubstake: IXO, Inc., literature and "Pocket Computer Terminal Said to Speed Access to Large Data

Banks," William M. Bulkeley (*WSJ,* February 25, 1982); "Toying with Computers," Bruce G. Posner (*INC,* September 1982); "You Can Take It with You," Stanley W. Angrist (*FBS,* December 6, 1982); " 'We Decided to Make a Million' " (*USNWR,* December 27, 1982/January 3, 1983); and "IXO: The Revolution Has Been Postponed" (*INC,* March 1983).

197 It's an amusing footnote: [55], p. 66.

198 At universities during the free-speech sixties: Andrew Pollack, "Playing Games Electronically" (*NYT,* December 24, 1981).

199 "a mistake. . . . I had just hired": Jerry Bowles, "*Video Games* Interview—Nolan Bushnell" (*Video Games,* August 1982). See also Jerry Bowles article "Silicon Valley Days" (*World,* No. 2, 1982).

199 To test it out: Aaron Latham, "Video Games Star Wars" (*NYT Magazine,* October 25, 1981).

200 convinced Warner Communications: Peter Bernstein, "Atari and the Video-Games Explosion" (*FTN,* July 27, 1981).

201 Activision, Inc.: Company literature and "The New 'Publishers' in Computer Software," Gary Slutsker (*VTR,* September 1980); "The Riches Behind Video Games" (*BW,* November 9, 1981); "Video-Game Firms Face Tough Christmas . . . ," Laura Landro (*WSJ,* September 29, 1983); and "A Fading Rose," Stanley W. Angrist (*FBS,* November 7, 1983).

202 Another more recent Silicon Valley: "Captains of Video Launch New Games," Dave Farrell (*VTR,* December 1981); "Imagic Scores in Video Games," Thomas C. Hayes (*NYT,* November 22, 1982); and "Creating Video Games That Score" (*BW,* April 4, 1983). Atari's copyright infringement suit against Imagic is outlined in "Atari Sues Imagic on Copyright Issue" (*NYT,* November 30, 1982); and "Imagic Settles Suit by Warner's Atari Unit" (*WSJ,* January 5, 1983).

203 The master of ceremonies: " 'To win the business game, do what you know,' " Nolan Bushnell (*INC,* August 1979); "Nolan Bushnell's New Act—Singing Robots," Peter W. Bernstein (*FTN,* July 27, 1981); and "It's All Play Money," Art Garcia (*Express,* March 1982).

204 Needless to say: Prospectus (November 11, 1981) for Pizza Time Theatre's public offering.

204 The only one given much hope: "High Profits from a Weird Pizza Combination," Gwen Kinkead (*FTN,* July 26, 1982); and "Pigging Out—ShowBiz Loses Pizzazz" (*FTN,* April 4, 1983).

204 The option that caused Brock: Telephone interview with Melvin A. Fechter, Aaron Fechter's father (March 31, 1983).

206 In March 1984: "The Pitfalls in Mixing Pizza and Video Games" (*BW,* March 12, 1984); "Pizza Time Files for Protection," Vartanig G. Vartan (*NYT,* March 29, 1984); and "Pizza Time Files for Court Protection . . . ," Victor F. Zonana (*WSJ,* March 30, 1984).

206 Lion and Compass: "Valley Eat" (*FBS,* October 24, 1983).

206 Androbot, Inc.: "Bushnell's Pet Robot Venture," Gary Slutsker and Jon Levine (*VTR,* November 1982); "Here Come the Robots," Philip

Faflick (*TM,* March 7, 1983); "FRED Goes Public" (*FTN,* July 25, 1983); and "Warner's Atari Unit to Market Products of Founder's Firm" (*WSJ,* January 11, 1984). Other companies Bushnell has spawned through his venture-capital holding company, Catalyst Technologies, are described in "The Pied Piper of Sunnyvale," Robert A. Mamis (*INC,* March 1983).

206 This trend: For an overview of the best microcomputer software firms specializing in games, see "Fun and Games and Profits," Lisa Miller Mesdag (*FTN,* May 2, 1983).

208 Sierra On-Line, Inc.: "New Stars, New Firmament," Kathleen E. Wiegner (*FBS,* May 24, 1982); and "A Bit of Old-Style Imagination Leads to a High-Tech Success," Carrie Dolan (*WSJ,* February 7, 1983).

208 Battles—actual campaigns: "Exec Strategic Simulations," Allan Tommervik (*Softalk,* July 1981); "Sweet Success at Strategic Simulations," David H. Ahl (*CC,* December 1981); and "Once Hooked, It's Hard to Get Away . . .' " (*San Francisco Chronicle,* May 1982).

209 Spinnaker's cofounders: "Software Firm Taps Market for Education," Richard A. Shaffer (*WSJ,* November 15, 1983).

209 But it's a California denizen: Lee Gomes, "Secrets of the Software Pirates" (*ESQ,* January 1982); "Key Software Writers Double as Media Stars in a Promotional Push," Susan Chace (*WSJ,* December 12, 1983).

210 George Lucas: "The Empire Strikes Back," Gerald Clarke (*TM,* May 19, 1980); "The Empire Pays Off," Stratford P. Sherman (FTN, October 6, 1980); "The Man Who Found the Ark," Tom Nicholson (*NSWK,* June 15, 1981); "But Can Hollywood Live Without George Lucas?" Aljean Harmetz (*NYT,* July 13, 1981); " 'I've Got to Get My Life Back Again,' " Gerald Clarke (*TM,* May 23, 1983); and the biography *Skywalking: The Life and Films of George Lucas,* Dale Pollack (New York: Harmony, 1983).

211 traditional mechanical movie wizardry: Conventional cinematic special effects, those for *Star Wars* in particular, are discussed in [20].

213 Mathematical Applications Group: Company literature and the articles "The Faster 3-D Way to Computerized Design" (*BW,* November 21, 1977); "A Cheaper Way to Make Slides" (*BW,* January 22, 1979); "The Creative Computers," Sharon Begley (*NSWK,* July 5, 1982); and "Animation Goes High-Tech," Eileen Gillooly (*VTR,* August 1983).

215 Cinetron Computer Systems: Telephone interview with Charlie Vaughn (April 8, 1983). See "Two's Company," Cathryn Jakobson and Abby Solomon (*INC,* May 1982).

216 Digital Effects, Inc.: Correspondence with Judson Rosebush and the profile of the firm in the April 13/27, 1981 issue of *The Harvard Newsletter on Computer Graphics.*

216 The keystone of John Whitney: Company press releases and the articles "From Fortran to Film," Jan Johnson (*DT,* August 1982); "Turning Dreams into Reality with Super Computers & Super Visions," John Lewell (*Computer Pictures,* January/February 1983); and "Next . . . Total Scene Simulation," Gene Youngblood (*Video Systems,* February 1983).

218 George Lucas has staffed: The sketch of Lucasfilm's computer development division was based on a presentation by the division's director, Dr. Ed Catmull, at Pratt Center's Computer Animation Seminar (October 1, 1982) in New York City; and "George Lucas Moves His Sci-Fi Genius to the Production Room" (*BW,* December 5, 1983).

8 Crunching Numbers and Processing Words: The Work-Station Visionaries

Page

222 A Booz•Allen & Hamilton study: The study sampled three hundred "knowledge workers" in fifteen representative large American companies over a period of one year. The study claims to be the most extensive survey ever undertaken of the output, working habits, and attitudes of corporate managers and professionals. See Booz•Allen brochure *Productivity* and the article "Who Needs the Office of the Future?," Harvey L. Poppel (*HBR,* November–December 1982).

222 In the one corner: Bob Davis, "Newcomers in Personal Computers Have Trouble Breaking into Market" (*WSJ,* February 18, 1983).

224 Charles A. Muench, Jr.: "The Man Who Never Wanted to Be President," Cathryn Jakobson (*INC,* September 1981); and "The Reluctant Entrepreneur," Jayne A. Pearl (*FBS,* June 7, 1982).

224 TeleVideo's Kyupin Philip Hwang: "The All-American Success Story of K. P. Hwang," Gene Bylinsky (*FTN,* May 18, 1981); and "Luck & Pluck—TeleVideo Ventures to Success," Julia G. Brown (*EB,* April 1983).

224 Even Charles I. "Chuck" Peddle: "Chuck Peddle: Chief Designer of the Victor 9000" (*Byte,* November 1982) and "Victor's Drive to Be No. 3 in the U.S." (*BW,* January 24, 1983).

225 It's particularly unlikely: "Victor Technologies' Sudden Success Has Turned into a Nightmare" (*BW,* November 14, 1983); and "Victor, 2 Units, Asks Protection in Chapter 11," Erik Larson (*WSJ,* February 8, 1984).

226 Engineering knowledge: The Vector Graphic story is based on a personal interview with cofounder Carole Ely (July 13, 1982). Also see "Next Stop Wall Street," Susan Benner (*INC,* March 1981); "Computer Coup" (*TM,* March 22, 1982); "The Entrepreneur Sees Herself as Manager," Eliza G. C. Collins (*HBR,* July–August 1982);

"Can Vector Do It Again?" Michael S. Malone (*NYT,* July 10, 1983); and "Head of Vector Resigns" (*NYT,* March 22, 1984).

227 At least for Lore Harp: "Merger by Marriage" (*Savvy,* October 1982).

227 Who is the youngest: Intertec company literature.

228 Martin Allen, a mechanical engineer: "The High Priests of Hi Tech" (*DBM,* March 1982); "A CAD Pioneer Whose Lead Is Fading" (*BW,* December 20, 1982); and [77], pp. 108–15.

230 "CAD/CAM is becoming as fundamental": "A CAD/CAM World?" Thomas G. Dolan (*BAR,* December 22, 1980).

232 For several months, Apollo Computer: "The Attractions of Starting a New Venture Prove Irresistible to Some Entrepreneurs," William Bulkeley (*WSJ,* June 9, 1981); "Apollo Computer: Why a Highflier Isn't Afraid of the Big Boys" (*BW,* July 30, 1984); and the prospectus for Apollo's initial public offering (March 3, 1983).

9 From Gutenberg to Galactica: The Space-Age Communicators

Page

236 Experts estimate: [77], p. 124.

237 As you read this: "Business Communications Challenges in the '80s," a white paper prepared by International Data Corp. (*FTN* ad supplement, April 14, 1983); and "Linking Computers to Help Managers Manage," Bro Uttal (*FTN,* December 26, 1983).

238 The premier manufacturer: "How Tiny Hayes Made All the Right Connections" (*BW,* August 1, 1983).

239 David Gold: "Debate Over Office 'Networks' " (*NYT,* August 15, 1982).

239 Ralph Ungermann: "The Second Time Around" (*VTR,* February 1980); "Corporate Pioneer Starts a New Firm," Michael S. Malone (*SJM,* February 25, 1980); "Ungermann-Bass Is Not a Typical Start-Up Company," Evelyn Richards (*Peninsula Times Tribune,* March 19, 1980); "Jogging Entrepreneur Is Now Running More for His Local-Networking Firm" (*DC,* September 1980); and "Ungermann-Bass Agrees to Acquire Amdax Corp.," Bruce Hoard (*CW,* January 31, 1983).

242 The two founders provided: Ungermann-Bass, Inc.'s, prospectus (May 10, 1983) for its initial public offering.

242 Dr. Robert M. Metcalfe: "Do These People Look Like Millionaires?" Edward O. Welles (*CalToday,* December 5, 1982); and "Will It Be a Yea or a Nay for Ethernet?" (*BW,* July 27, 1983).

243 When 3Com went public: Preliminary prospectus (January 25, 1984).

244 But the price is coming down: "Voice Mail Improves Its Delivery" (*BW,* December 13, 1982).

244 VMX, Inc.: " 'Three's the Charm,' " Anne Bagamery (*FBS,* March

29, 1982); and "From Trash to Cash," Craig R. Waters (*INC,* November 1983).

245 "the whole industry": "Going to Court to Protect Patents" (*VTR,* August 1983).

247 You could say that Vitalink: Jon Levine, "Living the Dream" (*VTR,* March 1983). Also, an explanatory letter (dated August 23, 1983) from Vivian E. Blomenkamp, a company founder.

247 It would be inaccurate: "Satellite Costs That Are Down to Earth" (*BW,* July 5, 1982); and "The Old Boy Satellite Network," Stephen Kindel and Jon Schriber (*FBS,* November 8, 1982).

10 Blue-Collar Automation: The Roboticians

Page

251 What is a robot?: The term "robot" is derived from the Czech word *robota,* meaning compulsory service. Karel Čapek, a Czechoslovakian playwright, coined the word "robot" to describe the mechanical men and women in his play *R.U.R.* (the initials stand for Rossum's Universal Robots). The play opened in London in 1921 and the word, henceforth, was used by science-fiction writers to describe most metallic forms of artificial life. In their novels, robots usually are vaguely man-shaped creatures that are very strong, single-minded, and dangerous.

For a layman's view of the history of robots as well as the technology's expanding horizons, read [44] and the excellent article "Industrial Robots on the Line," Robert Ayres and Steve Miller (*TR,* May/June 1982).

251 The Japanese, for example: For more on Japan's robot juggernaut, see "New in Japan: The Manless Factory," Steve Lohr (*NYT,* December 13, 1981); "The Push for Dominance in Robotics Gains Momentum" (*BW,* December 14, 1981); "The Robots Are Coming and Japan Leads Way," Hidehiro Tanakadate (*USNWR,* January 18, 1982); and "The Japanese Art of Automation Shown in Car Plants' Robots" (*NYT,* March 28, 1983).

251 automated machines: The word "automation" has its origins in the automobile industry and was coined by a Ford Motor Company engineer, Delmar S. Harder, to describe the system he devised in the 1940s to manufacture auto engines. Harder's completely automatic process reduced the time for producing each engine from twenty-one hours to fourteen minutes. It was the first totally self-regulating system applied to manufacturing.

In 1952 John Diebold, a well-known New York City management consultant, sanctioned and familiarized the word even further by titling his classic book on the subject *Automation.*

253 "As of today [1981]": "The Impacts of Robotics on the Workforce

and Workplace" (Pittsburgh: Department of Engineering and Public Policy, Carnegie-Mellon University, 1981), pp. 2–3.

253 Joseph F. Engelberger: "An Interview with Joseph Engelberger," Jerry W. Saveriano (*Robotics Age,* January/February 1981); and "Springtime for an Ugly Duckling," Subrata N. Chakravarty (*FBS,* April 27, 1981).

254 In a series of stories: Isaac Asimov describes the evolution of his thinking and writing about robots in [4], pp. 288–90.

255 "They seemed to have": John W. Dizard, "Giant Footsteps at Unimation's Back" (*FTN,* May 17, 1982).

256 "I was wrong": Barnaby J. Feder, "He Brought the Robot to Life," (*NYT,* March 21, 1982).

257 When Nissan Motor Company: John A. Byrne, "Whose Robots Are Winning?" (*FBS,* March 14, 1983).

257 A 1981 survey: Robert U. Ayres and Steven Miller, "Robotics, CAM, and Industrial Productivity" (*NPR,* Winter 1981–82).

257 Robots were seized upon: In 1981, the Department of Engineering and Public Policy of Carnegie-Mellon University surveyed the 124 member firms of the Robot Institute of America. One series of questions, directed at the corporate users or potential users of robots, concerned their motivations for adopting the new technology. The current users gave nine reasons, in this order: (1) reduced labor cost; (2) elimination of dangerous jobs; (3) increased output rate; (4) improved product quality; (5) increased product flexibility; (6) reduced materials waste; (7) compliance with OSHA regulations; (8) reduced labor turnover; and (9) reduced capital cost. The prospective users who had yet to purchase their first robot ranked their motivations in the same order, with three exceptions: They ranked "improved product quality" second; "elimination of dangerous jobs" third; and "increased output rate" fourth.

See "The Impacts of Robotics on the Workforce and Workplace," op. cit., pp. 47–53.

258 Cincinnati Milacron: "Promising," Maurice Barnfather (*FBS,* August 16, 1982); "Cincinnati Milacron, Mainly a Metal-Bender, Now Is a Robot Maker," Paul Ingrassia and Damon Darlin (*WSJ,* April 7, 1983); and "Comes the Revolution!" James Cook (*FBS,* October 24, 1983).

258 PRAB Robots, Inc.: Company literature and the articles "PRAB Robots, Inc., Keeps It Simple" (*INC,* June 1982); and "An Interview with Walter Weisel," Edgar T. Pilson (*Travelhost,* July 11, 1982).

259 When PRAB went public: The prospectus (September 17, 1981) for the initial public offering.

260 Advanced Robotics Corporation: The information on the company was supplied by Ronald C. Reeve in two letters.

261 Artificial vision: [22], pp. 117–18.

261 John W. Artley: The Object Recognition Systems sketch is drawn from literature supplied by the company as well as the article "Auto-

mated Inspection: The New Technology," John W. Artley (*Design News,* July 5, 1982).

263 Machines Intelligence, Inc.: Jerry Buckley, "Venturers Bet on Artificial Intelligence" (*VTR,* March 1981).

263 Cognex Corporation: "Offices of the Future," Kay Dockins Ingle and Sharon Pavlista (*VTR,* June 1982); "OCR System Solves Automation Problems," Eric Lundquist (*MMS,* August 1982); and "Now—A New Way to Scan Man-Readable Data" (*MMH,* August 6, 1982).

264 "Nineteen eighty-two was a transition year": "Eyes for the Automated Factory" (*HT,* April 1983).

264 In the early 1980s: "Now Everybody Wants to Get into Robots," (*BW,* February 15, 1981); and "Hope for Robot Makers After a Flat Year," Kevin Farrell (*VTR,* June 1983).

264 But the circus: Kevin Farrell, "A Fast-Approaching Shakeout in Robots" (*VTR,* June 1982).

265 Westinghouse Electric Co.: "Westinghouse Signs Definitive Agreement to Buy Unit of Condec" (*WSJ,* January 10, 1983); and "Unimation's 'Father of Robotics' Viewed as Big Asset in Westinghouse Purchase," Doron P. Levin (*WSJ,* January 12, 1983).

266 "trying to run": Carol Hymowitz, "Manufacturers Press Automating to Survive, but Results Are Mixed" (*WSJ,* April 11, 1983).

266 Phoenix Digital Corporation: Corporate literature and the articles " 'Some Things You Have to Experience,' " Lisa Gross (*FBS,* December 7, 1981); and "Factory Computers Link Up" (*BW,* July 5, 1982).

268 Control Automation: Jerry Buckley, "Venturers Bet on Artificial Intelligence" (*VTR,* March 1981).

268 Philippe Villers: "The Attractions of Starting a New Venture . . . ," William Bulkeley (*WSJ,* June 9, 1981); "A New Foundation Will Aid the Aged," Kathleen Teltsch (*NYT,* August 15, 1982); "Computervision Says Ex-Officer Is Cleared of Stealing Information" (*WSJ,* November 8, 1982); and the prospectus (March 1, 1983) for Automatix's initial public offering.

268 "lean and mean": Michael Ball, "Automatix Runs 'Lean' on Information Software" (*EB,* April 1983).

269 "In our view": "The Advent of Turnkey Robotic Systems" (*RT,* Fall 1980).

270 In manufacturing terms: Alvin Toffler describes in detail this trend toward shorter production runs in [75], Chap. 15.

271 "By 2010 or so": Robert U. Ayres and Steven M. Miller, "Robotics and Conservation of Human Resources" (Pittsburgh: Department of Engineering and Public Policy, Carnegie-Mellon University, 1982), p. 6.

271 According to Carnegie-Mellon's: Alix M. Freedman, "Behind Every Successful Robot Is a Technician" (*NYT,* October 17, 1982).

271 "At first I was bitter": Joann S. Lublin, "As Robot Age Arrives, Labor Seeks Protection Against Loss of Work" (*WSJ,* October 26, 1981).

Index

Abel, Robert, 215
Abel, Robert, and Associates, 215
Activision, 201–202
Adler, Frederick, 268
Advanced Micro Devices (AMD), 111–114
Advanced Robotics, 260
Agrigenetics, 74
Ahl, David, 102
aircraft manufacturing, 42
AirFone, 176–177
Airsignal, 176
Alcorn, Al, 199
Allen, Martin, 228–230, 269
Allen, Paul G., 154–155
Allied Corporation, 46
Amdahl, 94–95, 117
Amdahl, Carlton, 94, 95, 96
Amdahl, Gene M., 94–95, 117
American Express, 176
American Fiber Optics (AMFOX), 187–188
American Research and Development, 41
American Telephone and Telegraph. *See* AT&T
Ameritech, 174
Ampex, 198
Anderson, Harlan, 41
Androbot, 206
antibodies, synthesis of, 70–71
antigens, 70
Apollo Computer, 232
Apple Computer, 44, 77–82, 83–88, 92, 101, 105, 128–129, 131, 153, 155, 206, 242
Apple Lisa, 101, 128–129, 136–137, 146, 222
Apple Macintosh, 128, 136
Apple I, 78
Apple II, 79–80, 82, 84, 87, 88, 128, 130, 132, 217, 225–226
Architecture Machine Group, 210
arc welding, 260–261
Arista Management Systems, 144
Artel Communications, 188
Arthur Rock & Associates, 84
artificial intelligence, 251, 263
Artificial Intelligence Center, 263
artificial vision, 261–264
Artley, John W., 261, 262
Asimov, Isaac, 254–255
ASK Computer Systems, 147–150, 268
Aslami, Mohd A., 188
Association of Data Processing Service Organizations, 142
AT&T (American Telephone and Telegraph), 39, 159–170, 172, 178, 179, 184, 187, 235
AT&T Communications, 160
AT&T Information Systems, 160
AT&T International, 160
AT&T Long Lines Division, 160, 167, 169
AT&T Technologies, 160
Atari, 78, 194, 199–201, 202, 203
automated machines, 251–252
Automated Welding, 260
Automatic Data Processing (ADP), 142–143
automation, 221–271
Automatix, 268–269
automobile industry, 27, 30, 31, 86, 88–89, 256–257, 263
AUTRAX, 179
Aweida, Jesse, 100, 123–124, 125
Ayres, Robert, 271

Baer, Ralph, 185
Baker, William, 134
bank teller machines, automated, 44, 98
baseband networks, 238
BASIC, 154, 155

Basic Systems, 73
Bass, Charlie C., 240
Battlezone, 194
Beck, Bob, 92
Beckman Instruments, 43
beepers, 174–176
Bell, Alexander Graham, 159–160, 184, 185
Bell Laboratories, 40, 41, 140, 160, 161, 187
B. F. Goodrich, 145
Billings, Joel, 208
Biogen, 65–67
biotechnology, 53–74
Birndorf, Howard C., 72
"blue box," 78
Boeing 247, 42
Boeing 707, 42
Bonneville Power Administration, 155
Booz · Allen & Hamilton, 222
Bosomworth, Ken, 194
Bovoroy, Roger, 116
Bowman, William, 209
Bowmar Brain, 44
Bowmar Instrument Corp., 44, 45
Boyer, Herbert W., 38, 54, 55, 58–60, 63–64
Bradley, Milton, 195
Bradley, Terry, 208
Breakout, 200
Bricklin, Daniel, 131–132, 133
broadband networks, 238
Brock, Robert, 204
Brock Hotel Corporation, 204
Bruno, Albert V., 45–46
Budge, Bill, 209
BudgeCo, 209
Buick, David Dunbar, 89
Bureau of Labor Statistics, 36
Bushnell, Nolan K., 78, 193–194, 197–200, 202–204, 205–206
Byers, Brook H., 71–73
Byte, 129–130

Cable and Wireless, 163
CAD (computer-aided design), 110, 215, 228–233, 265, 266, 268
CAE (computer-aided engineering), 231
calculators, 84, 105
CAM (computer-aided manufacturing), 230, 266, 269
Campbell, Gordon A., 116
Cape, Ronald E., 56, 58
capitalist economy, cycles in, 27–31, 88–90
Carlson, Chester, 42–43
Carnegie-Mellon University, 252–253, 271
Carter, Jimmy, 36–37
Carter, Thomas F., 161–163
Carterfone, 162–163, 165
Cassandra list, 45
Catalyst Technologies, 206
cell fusion, 71
cellular telephony, 170–174
Centel, 174
Center for Biotechnology Research, 70
Centocor, 72
Cerny, Richard, 188
Cetus Corporation, 55, 56–58, 60, 65, 74
Charney, Howard, 242
Chrysler, 89
CIF (computer-integrated factory), 266
CIM (computer-integrated manufacturing), 265–271
Cinetron Computer Systems, 215–216
Citibank, 178
coaxial cable system, 186
COBOL, 155
Cogar, 44–45
Cognex, 263–264
Cohen, Fred, 178, 179
Cohen, Stanley N., 54, 55, 58, 63–64
Combined Network, 169
Commodore PET, 225–226
Communications Industries, 176
computer graphics, 215
ComputerLand, 82–83, 208
Computer Pictures Corporation, 146
computers, 77–105
Computer Space, 198, 199
Computer Store, The, 82
computer stores, 82–83, 133
Computer-Tabulating-Recording Company (C-T-R), 90
Computervision, 228–230, 268, 269
Concord Data Systems, 237
Conigliaro, Laura, 264
Conner, Finis F., 117–121
Consolidated Controls, 255
Context Management Systems, 137
Control Automation, 268
Control Data, 92–93, 94, 124
Coolidge, Calvin, 38, 39
Cooper, Arnold C., 45–46
Coppola, Francis Ford, 213
copycat companies, 110
Corning Glass, 186, 187, 188–189
CP/M operating system, 151, 155
CP/M-86 operating system, 153
CPUs (central processing units), 116–117, 118
Crane, David, 201–202
Cray, Seymour R., 93
Cray-1, 217
Cray Research, 93–94

"creative destruction," 30
Creative Engineering, 204
Crick, Francis H. C., 53
cross-breeding, 73
Crowder, Dony, 178
CRTs (cathode ray tubes), 125
Cullinane, John J., 145–146
Cullinet Software, 145
CULPRIT, 145
Curry, Eddie, 77
CXC, 184
Cytogen, 72

Dabby, Frank W., 188
daisy-wheel printers, 126
data bases, 83, 234
data communications, 236–237
Data General, 182, 268
Data General NOVA, 182
Datamation, 132
DDP (distributed data processing), 236–237
Death Star, 209
DEC (Digital Equipment Corporation), 41–42
"Decline of Productivity and the Resultant Loss of U.S. World Economic and Political Leadership, The," 257
Delbruck, Max, 53, 54
Deleage, Jean, 123
Demos, Gary, 216–218
Devol, George, 255
Diablo Systems, 126, 127
Digital Effects, 216
Digital Equipment Corporation. *See* DEC
Digital Productions, 216–218
Digital Research, 151–153
Digital Research CP/M operating system, 151, 155
Digital Research CP/M-86 operating system, 153
Dion, C. Norman, 120–121
Disney, Walt, 193, 194
Disney, Walt, Productions, 213, 216
DNA (deoxyribonucleic acid), 53–54, 58, 62, 63, 70, 73
Docutel, 44, 45
Dodge, Frank, 146, 147
dot-matrix printers, 125–126
double helix, 53–54
Dougherty, Brian, 202
Douglas DC-2, 42
Douglas DC-8, 42
Doyle, Robert and Holly, 195–196
Dun & Bradstreet, 45, 146
Durant, William C., 89
Duryea, Charles and Frank, 31
Dysan Corporation, 120–121

EEPROM (electrically erasable, programmable read-only memory), 116
802 standard network, 239
electrical engineers, 46
Electronic Data Systems (EDS), 141–142
electronic mail, 222
Eli Lilly, 62
Ely, Carole, 226–227
embedded data processing, 237
Engelberger, Joseph F., 253–256
Engineering Research Associates, 92
entrepreneurship, 36–37, 38, 42, 46–47, 104–105
Environmental Protection Agency (EPA), 257
Ethernet, 239, 242
Eugenics, 70
Evans, David C., 197
Evans & Sutherland, 197, 216
Evans & Sutherland LDS-1, 197
Exxon Enterprises, 240

Faber, Edward, 83
factories, automated, 251–271
Faggin, Federico, 240
Fairchild Semiconductor, 41, 43, 111, 112, 115
Farley, Peter J., 56, 58
Fechter, Aaron, 204, 205
Federal Communications Commission (FCC), 160, 167, 168–169, 173, 175
fertilization, crop, 73–74
Fiber Communications, 188
fiber optics, 170, 184–189, 246, 267
film industry, 209–218
Fischer, Brian, 137
floppy disks, 80, 117, 118
Florist Transworld Delivery system, 163
Fogarty, James, 173
Food and Drug Administration (FDA), 55, 57, 62, 72
Ford, Henry, Sr., 27, 31, 77, 85–86, 103, 159, 270
Ford, Henry, II, 74
Ford Model T, 31, 85–86, 270
Ford Motor Company, 27, 86, 89
FORTRAN, 132, 155
Fortune 500, 40, 57, 85, 98, 106, 124, 179, 223, 240
Fortune 1,000, 37
Foster, William E., 99
"foundries," 110
Fox, Arthur, 263
Frankston, Robert, 132, 133
Freiburghouse, Robert A., 99
Fujitsu, 95
Future Computing, 105

Fylstra, Daniel H., 129–134, 135–136, 138, 140

Gates, William H., III, 154–156
Gebelli, Nasir, 208–209
Gebelli Software, 209
"gene machine," 59
Genentech, 58–60, 61, 62–64, 65, 71, 74
General Electric, 84, 265
General Motors, 89, 142, 256, 265, 267–268
General Telephone and Electronics (GTE), 162
gene-splicing, 54–55, 63–64
genetic engineering, 53–74
Genetic Prophecy (Harsanyi), 64
Genetic Systems, 72
Genex, 65, 66
Gilbert, Walter, 65, 67, 68, 69
Glaser, Donald A., 56, 58
Glass, Marvin, & Associates, 195
Glick, J. Leslie, 65, 66
Goeken, John D., 163–164, 165, 176–177
Goell, Jim, 188
Gold, David, 239
Goldberger, Jim, 202
Goldstein, Richard, 67–68
Goodrich, B. F., 145
"graded index" optical fiber, 188
Grand Cayman Island, as tax haven, 61–62
Graphic Scanning, 173–174, 175–176
Graves, William M., 144, 145
Green, Michael D., 98, 99
Grimm, W. T., 144
Grove, Andrew S., 44, 106, 107, 110
growth hormones, 60, 62
Grubb, William F. X., 202

Haloid Co., 43
hand-held computers, 105
Harp, Lore, 226–227
Harp, Robert, 226, 227
Harsanyi, Zsolt, 64
Harvard Business School, 131–132, 149–150
Harvard University, research guidelines of, 67–69
Hayes, Dennis C., 238
Hayes Microcomputer Products, 238
Heiser, Dick, 82
Helling, Robert B., 64
Hendrie, Gardner C., 99
Hewlett, William R., 97
Hewlett-Packard, 78, 96, 97, 148
Highland Park Factory, 27
Hollerith, Herman, 90
holography, 210
Horley, Albert L., 247
Hoxie, Gib, 137
Hudgins, Don, 215
Hughes Communications, 170
Humulin, 62, 63
Hwang, George, 97
Hwang, Kyupin Philip, 224
hybridization, 73
hybridoma technology, 71–72
Hybritech, 71–73

Ibis Systems, 124
IBM (International Business Machines), 39, 41, 43, 85, 90, 91, 94, 105, 110, 117, 123–124, 141, 143–147, 153, 184, 230, 235, 239
IBM Model 701, 91
IBM Model 1401, 80–81
IBM PC, 105, 139–140, 227, 242
ICs (integrated circuits), 41, 43, 106, 110, 114
Imlay, John P., Jr., 144, 145
IMSAI, 77, 82
Industrial Light and Magic, 218
Information International, 216
Information Unlimited Software, 134
InnoVen, 65
Institute of Electrical and Electronic Engineers (IEEE), 108
insulin, synthesis of, 60, 62
InteCom, 184
Integrated Data Management System (IDMS), 145–146
Intel, 43–44, 45, 106–110, 115, 151
Intel 432 microminiaturized mainframe, 108–109
Intel 4004 microprocessor, 44, 107, 109
Intel 8008 microprocessor, 44, 107, 108, 109
Intel 8086 microprocessor, 108
Intelligent Systems, 223–224
interconnect industry, 162
interferon, 60, 62
Intermetrics, 129
International Business Machines. See IBM
International Data, 227, 237
International Memories, 117
International Plant Research Institute (IPRI), 74
International Resource Development, 194
International Telephone and Telegraph (ITT), 127
Intertec Data Systems, 227–228
IXO, 196

Jaeger, Raymond E., 188
Japanese competition, 64–65, 105, 112, 124–126, 135, 223, 227, 251–252, 256–257, 265, 266

Jefferson, Thomas, 38
Jennings, Peter, 131, 135
Jobs, Steven P., 38, 77–81, 87, 101, 102, 225–226
Johnston, Robert F., 65, 66
Juliussen, Egil, 105, 223
Justice Department, 39, 161, 179

Kaplan, Larry, 201–202
Kapor, Mitchell D., 138–140
Katzman, James A., 98, 99
Kawasaki Heavy Industries, 256–257
Keenan, Joseph F., 199, 200, 202, 204
Kelly, Eugene W., 145
Khashoggi, Adnan, 183
Kidde, 225
Kilby, Jack, 43*n*
Kildall, Gary and Dorothy, 151–153
Kleiner, Perkins, Caufield & Byers, 58–60, 73, 115, 116, 202, 247
Kleist, Robert A., 125–126
Koble, Dennis, 202
Kohler, Georges, 70–71
Kondratieff, Nikolai Dimitriyevich, 28
Kondratieff wave, 27–31
Koppers, 65
Kubrick, Stanley, 216
Kurtzig, Arie, 148–149
Kurtzig, Sandra, 146–150, 268

Land, Edwin, 43
Landry, C. Kevin, 65–67, 68, 123
LANs (local area networks), 221–222, 236–243, 246
Larson, Charles, 258
lasers, 186
Lautenberg, Frank, 143
Lawrence Livermore Laboratory, 231
LED (light-emitting diode), 186
Lee, David, 126–127
Lembas, Gerald A., 123
Leslie, Mark, 100
Levy, James H., 202
lightwave communication, 170, 184–189
Lightwave Technologies, 188
Lilly, Eli, 62
Lion and Compass restaurant, 206
Lisberger, Steven, 213
Loewenstern, Walter, Jr., 180–184
logic ICs, 108
"long wave" economic theories, 27–31
Lotus Development, 139, 140
Lotus 1-2-3 software program, 138–140
Loustanou, John C., 98, 99
Lubrizol, 74
Lucas, George, 210–213, 218
Lucasfilm, 210
Lundin, Leonard B., 123

McCormack, James, 146, 147
McCormack & Dodge, 146, 147
McDonald's Systems, 106
McDonnell Douglas, 42
McGovern, Patrick, 227
McGowan, William G., 164–170, 176, 178
Machine Intelligence, 263
machine-tool industry, 253
Madden, Clifford J., 95
Magna Theatre Corporation, 164
Magnavox, 195
Magnetic Peripheral Inc. (MPI), 121–122
magnetic recording heads, 121–123
Magnuson Computer Systems, 96
mail, electronic, 243–245
mainframe computers, 81, 84, 86, 89, 90–96, 140–141, 144, 182, 230
Management Science America (MSA), 143–145
ManMan Information System, 148, 268
Mansfield, E. Blaine, Jr., 187
Marconi, Marchese Guglielmo, 185
Mark, Hans, 33
Mark I, 86
Markkula, A. C., Jr., 38, 79, 84–85, 88
Marvin Glass & Associates, 195
Massachusetts Institute of Technology (MIT), 28, 210
Math, Irwin, 189
Math Associates, 189
Mathematical Applications Group, Inc. (MAGI), 213–214, 215
Mattel Intellivision, 202
Matthews, Gordon, 244, 245
Maxfield, Robert R., 180–184
MBA software program, 137
Melchor, Jack, 182
Memorex, 124
memory chips, 115, 117
Mensch, Gerhard, 30–31
Merlin, 195–196
Merrill Lynch, 84, 187
Meshulach, Avraham, 47
Metcalfe, Robert M., 242–243
Metromedia, 176
MicroChess, 131, 132
microcomputers, 44, 77, 78, 85, 86, 90, 101–105, 108, 121, 127, 129, 206
Micro Instrumentation and Telemetry Systems. *See* MITS
microprocessors, 78, 107–110
MicroPro International, 134
Microsoft, 153–156

Microsoft MS/DOS operating system, 153, 156
Microwave Communications, Inc. (MCI), 163–170
microwave transmission, 167–168, 246
Milacron, 258
military games, 194–195
Millard, William H., 77, 82–83
Miller, Alan, 201–202
Miller, Ken, 237
mil-spec computers, 182–183
Milstein, Cesar, 70–71
Milton Bradley, 195
minicomputers, 41, 89, 96–100, 232–233
MITS (Micro Instrumentation and Telemetry Systems), 77, 155
MITS Altair, 77, 155
Mittelman, Phillip S., 213–214
MK-60, 194
mobile phone service, 170–174
modems, 237–238
monoclonal antibodies, 71–72
Monoclonal Antibodies, Inc., 72
"mono mode" optical fiber, 188
Moore, Gordon E., 43, 107, 108, 110, 112
Moore's Law, 108
MOS 6502 microprocessor, 225
Mostek, 139
Motorola, 175
Mott, Stewart, 74
"mouse," 136, 137
movie map, 210
MS/DOS operating system, 163–170
Muench, Charles A., Jr., 224
MUSE Software, 209
Mystery House, 207

Naisbitt, John, 221
National Geographic, 159
National Institutes of Health, 55
National Semiconductor, 46
Nature, 53
Naugler, W. Edward, Jr., 187
Nestle, Elliot, 100
Net/One, 240
New York Telephone, 171
New York Times, 63, 115
Nissan Motor Company, 257, 268
nitrogen fertilizers, 73
NonStop system, 98
Norris, William C., 92–93
Norrsken, 188
North American Telephone Association, 163
Nottingham, Gene, 215
Noyce, Robert N., 43–44, 78, 107, 108, 110, 112, 115
nuclear energy, 39
Nutting Associates, 199

Object Recognition Systems. *See* ORS
Occupational Safety and Health Administration (OSHA), 257
Odyssey, 195, 197
OEMs (original equipment manufacturers), 119, 124–125, 223, 260
offices, automated, 221–248
Olsen, Kenneth H., 41, 42
O'Neill, Gerard, 31–33
1-2-3 software program, 138–140
optical fibers, 184–185
optical laser memory device, 124
ORS (Object Recognition Systems), 261–263
ORS Ibot-1, 262
Osborne, Adam, 102–104
Osborne Computer, 102–105
Osborne Executive 2, 104
Osborne 1, 102, 103
Oshman, M. Kenneth, 180–184
Otto, Nikolaus, 31

Pacific Technology Venture Fund, 227
Packard, David, 97
Packard-Bell, 92
Padwa, David, 73–74
PageAmerican Group, 176
PageGram, 176
paging services, 174–176
Palevsky, Max, 74, 90–92
Parker Brothers, 195
Pascal, 155
Patent and Trademark Office, U.S., 55, 64
patents, 55, 63–64, 70
PBX (private branch exchange), 180, 183–184, 238, 244
PCMs (plug-compatible manufacturers), 94
Peachtree Software, 134, 144
Pearlman, Jerry, 38
Pearson, Hal, 216
Peddle, Charles I., 102, 224–226
Penn Central Railroad, 27
Perceptron, 263
Perfect Software, 134
peripheral equipment, 116–124, 150
Perot, H. Ross, 141–142
personal computers, 44, 80
Personal Software. *See* VisiCorp
Pertec Computer, 123, 125
Petrozzo, Michael, 178
Phoenix Digital, 266–268
photophone, 185
Pizza Time Theatre, 193, 202–204, 205–206
planar process, 43*n*

PL/M language, 151
Poduska, J. William, 232–233
Polaroid, 43
Pong, 199, 200
portable computers, 102–105
Post, David, 176
PRAB Robots, 258–260
"Preliminary Discussion of the Logical Design of an Electronic Computing Instrument" (Von Neumann et al.), 140
Prime Computer, 232
Printronix, 126
protocols, 236
public offerings, 57, 61, 62, 135
punch-card calculating systems, 90

Qantel, 96
Qume, 126–127

Radio Corporation of America. *See* RCA
Radio Shack, 153
radiotelephone services, 170–177
railroad industry, 27
Ramtek, 218
Randall, Eric N., 188–189
R&D (research and development), 36, 39–40, 43, 55, 101, 109
Randolph, A. Philip, 30
Raster Blaster, 209
RCA (Radio Corporation of America), 41
RCA Global Communications, 176
RCCs (radio common carriers), 172, 174, 175
Reagan, Ronald, economic policies of, 36–37, 38–40
Reeve, Ronald C., Jr., 260
Richeson, Gene, 180–184
Riggs, Henry E., 40
Robert Abel and Associates, 215
Roberts, Ed, 77
Robertson, Gordon I., 268
robotics, 203–206, 230, 251–271
Robot Institute of America, 252, 257
Rock, Arthur, 92, 106, 135
Rock, Arthur, & Associates, 84
Rockefeller, John D., 159
Rodde, Anton F., 268
Rollins, Paul F., 266–268
ROLM, 180–184
ROLM CBX, 183–184
ROLM CBX II, 184
Roosevelt, Franklin D., 30
Rosebush, Judson, 216
Rosen, Benjamin, 139
Rosen, Charles, 263
Royston, Ivor, 72
Rubinstein, Seymour, 134
Runyon, Damon, 138
Russell, Steve, 198

Salsbury, Phillip J., 116
Sanders, W. J., III, 111–114, 323
Sanders Associates, 195
Satellite Business Systems (SBS), 235
satellites, communication, 246–248
Schafler, Norman, 255
Schroeder, Robert E., 126–127
Schumpeter, Joseph A., 30
Scientific Data Systems (SDS), 92
Seagate Technology, 118–121, 126
"second sourcing," 112
Securities and Exchange Commission, 37
SEEQ Technology, 114–116
semiconductor industry, 106–116
Service Bureau Corporation, 141
Seuss, David, 209
Sevin, L. J., 139
Sevin Rosen Partners, 139, 140
Shaw, Carol, 202
Shillman, Robert J., 264
Shockley, William, 43
Shockley Semiconductor Laboratories, 43
ShowBiz Pizza Place, 204, 205
Shugart, Alan F., 47, 117–121
Shugart Associates, 117, 118
Siemens, 113–114
Sierra On-Line, 206–208
Silicon Valley, 45, 97
Simon, 195
Sinclair, Clive, 103
Singleton, Henry, 84
Sirius Software, 208, 209
Skywalker Ranch, 211
Small Business Administration, 37
small businesses, 36–37, 45
Smithsonian Institution, 36, 228–229
Softporn, 208
software, 83, 105, 109, 128–156
Software Arts, 132, 133
Software Publishing, 134
solid-state electronics, 40
somatostatin, synthesis of, 60
Sorcim, 134
Southern Biotech, 61–62
Space Invaders, 201
Space Wars, 198
specialized common carriers, 167
SpecTran, 188
speech synthesis, 210
Sperry Univac, 183
Spielberg, Steven, 213
Spinnaker Software, 209

Sports-Pac program, 96
Sprague, Peter, 46
stand-alone programs, 137
Stanford University, 63–64, 69–70
Star Wars, 211, 212, 213
Storage Technology, 123–124, 125
Strategic Simulations, Inc. (SSI), 208
Stratus Computer, 99–100
Studebaker, Clement, 89
Stutz, Harry C., 89
Susnjara, Ken J., 260
Sutherland, Ivan E., 197
Sutter Hill Ventures, 126, 202, 232
Swanson, Robert A., 38, 58–60, 61
Synapse Computer, 100
Synapse N+1 computer, 100
SynthaVision, 214–215
System Dynamics National Model, 28
Syzygy, 199

TA Associates, 66, 67, 208, 209
TANDEM Assay System Kits, 72
Tandem Computers, 98–99, 100, 247
Tandon, Sirjang Lal, 121, 125
Tandon Corporation, 121–123
telecommunications industry, 33, 159–189, 234–248
TelePrompter Manhattan Cable Television, 187
telequipment, 178–189
TeleSciences, 178–180
Teleswitcher, 245
TeleVideo Systems, 223–224
Texas Instruments, 43, 107, 116, 161
Thermwood, 260
Thorpe, Jack, 194
3Com Corporation, 242–243
3Com I.E. controller, 242
"Three Laws of Robotics" (Asimov), 255
Threshold Technology, 245
Todd, Mike, 164
Todd-AO process, 164
Toffler, Alvin, 221
"total systems solution," 143
Traf-O-Data, 155
transistors, 41, 43, 161
Traub, Henry, 143
Treybig, James G., 97, 98, 99, 247
Trilogy, 94–96
Trilogy Systems, 40
Trombly, John, 263
TRON, 213
turnkey system, 148, 230
2001—A Space Odyssey, 216
2081 (O'Neill), 31–33
Tymshare, 148

Ultrasonic, 164, 165
unemployment, in economic cycles, 28–31
Ungermann, Ralph, 239–242
Ungermann-Bass, 240–242
Unimation, 254–256, 268
United States Census of Manufacturers, 88
United States Robots, 264
universities, research work in, 67–70
University of California at Berkeley, 69–70
uptime programs, 100
US Festival, 88
U.S. Telephone Communications, 169

Valentine, Donald T., 79
Valid Logic Systems, 231
Vanderbilt, Cornelius, 27
Vaughn, Charlie, 215
Vazirani, Hargovind N., 187
VCS (video computer system), 200–201, 202
Vector Graphic, 77, 226–227
Venrock Associates, 84
venture-capital resources, 37–38, 41, 47, 61, 71, 112–116, 121, 127, 134–135, 153–154, 223–224, 232
Viatron, 44–45
Victor Technologies, 224
videodisks, 117–121, 210
video games, 193–209, 224
video monitors, 125, 223–224
Villers, Philippe, 268–269
VisiCalc, 83–84, 128, 130, 132, 133, 137, 140
VisiCorp, 128–137, 138, 139, 140
VisiON, 128–129, 132, 136–137
vision, artificial, 261–264
Vitalink Communications, 247–248
VLSI (very-large-scale integrated) circuits, 231
VMX (Voice Message Exchange), 244–245
voice-mail, 243–245
Von Neumann, John, 140

Wallace, John, 258
Walt Disney Productions, 213, 216
Warner, Silas, 209
Warner Communications, 200, 202, 203
Warp Factor, 208
Washington, George, 229
WATSBox, 245
WATS line resellers, 169
Watson, Arthur K., 90
Watson, James D., 63
Watson, Thomas J., Sr., 90, 91
Watson, Thomas J., Jr., 90
WCCs (wire-line common carriers), 172, 174
Weinberg, Sidney, 86

Weisel, Walter K., 259, 260
Weiss, Mitchell, 264
Wells, William M., III, 227–228
Western Electric, 160, 179, 187
Western Union, 178
Weyerhaeuser, William, 74
Whitehead, Bob, 201–202
White House Conference on Small Business, 37
Whitney, John, Sr., 216
Whitney, John, Jr., 216–218
Williams, Kenneth and Roberta, 207–208
Winchester disks, 117–121
Winton, Alexander, 89
Wistar Institute of Anatomy and Biology, 70, 72
Witkowicz, Tadeusz, 188
Wizard and the Princess, 208
word processing, 127, 134
WordStar, 134
work stations, 221–233, 265
Wozniak, Stephen G., 38, 77–81, 87–88, 97, 225–226
Wright, Paul, 266
W. T. Grimm, 144
WUI, 170

xerography, 42–43
Xerox, 43, 84, 92, 117, 127
Xerox Star, 222

Yampol, Barry, 173–174, 175–176
Yaskawa, 263

Zaron, Ed, 209
Zilog, 240, 241